Inspiring Designs and Vibrant Plants
for the Waterwise Gardener

HOT COLOR, DRY GARDEN

Timber Press · Portland, Oregon

NAN STERMAN

This book is dedicated to my husband, Curt Wittenberg—my sweetie, my life partner, and supporter of all my dreams; to Gabe, Asher, and Tamar, the up-and-coming gardeners of the world; to Grandpa Eddie, who never got to be a farmer; and to Jan Smithen, my dear friend and inspiration.

Copyright © 2018 by Nan Sterman. All rights reserved.
Photography credits appear on page 307.

Published in 2018 by Timber Press, Inc.

The Haseltine Building
133 S.W. Second Avenue, Suite 450
Portland, Oregon 97204-3527
timberpress.com

Printed in China on paper from responsible sources
Fourth printing 2022

Text design by Michelle T. Owen
Cover design by Sarah Crumb and Hillary Caudle

Library of Congress Cataloging-in-Publication Data

Names: Sterman, Nan, author.
Title: Hot color, dry garden: inspiring designs and vibrant plants for the waterwise gardener / Nan Sterman.
Description: Portland, Oregon: Timber Press, 2018. | Includes bibliographical references and index.
Identifiers: LCCN 2017046300 (print) | LCCN 2017049590 (ebook) | ISBN 9781604698633 | ISBN 9781604694574 (pbk.)
Subjects: LCSH: Drought-tolerant plants. | Xeriscaping.
Classification: LCC SB439.8 (ebook) | LCC SB439.8 .S747 2018 (print) | DDC 635.9/525—dc23
LC record available at https://lccn.loc.gov/2017046300

A catalog record for this book is also available from the British Library.

Contents

Acknowledgments 6
Introduction 7

Using Color in the Garden 13
Designing for Plant Structure and Texture 38
Garden Gallery 48
Plant Directory 175
Dry Gardening How-To 284

Resources 303
Photography Credits 307
Index 308

ACKNOWLEDGMENTS

Thanks to designers Scott Calhoun, Judith Phillips, Patrick Anderson (yes, Patrick, you are a designer), Nick Wilkinson, Gabriel Frank, Nick Dean, Dustin Gimbel, Leslie K. Dean, Judy M. Horton, Amelia Lima, Laura Morton, and Alan Richards for creating the beautiful, color-filled, waterwise gardens that inspired this book and will inspire its readers. Thanks also to Santa Rosa landscape designer Michelle Bellefeuille, who passed on during the production of this book (see her garden on page 58). Michelle was one of the most gifted and thoughtful designers I've ever met. She was soft spoken and quiet, but with an aura that drew people to her—simply put, she glowed. Being around her was magical and I can't imagine not seeing her again. I hope featuring her gardens on these pages serves as a tribute to her accomplishments and a lasting memory of her lovely presence.

Thank you to all the designers and garden owners who allowed me into their private spaces to photograph and share their gardens. Thanks also to Glenda Garmon, Andrea Testa-Voight, Cheryl K. Lerner, Carol Anne McDougal, Dinah Grisdale, Macon McCrossen, Annette Gutierrez of Potted, Debra Carl, Scott Spencer, and Maggie Judge. Your generosity paves the way for others who are contemplating this important shift from thirsty, wasteful gardens to waterwise, resource-thrifty, and beautiful, colorful gardens.

Thanks also to the designers and homeowners who allowed me to shoot gardens that unfortunately ended up on the cutting room floor. With more gardens than pages, deciding which to feature and which to leave out was a tremendously painful process. Please be assured that your efforts were not wasted. Your gardens serve as examples in the many talks and presentations I give.

Thank you to all the garden designers, horticulturists, and gardeners who reviewed plant lists and shared their plant knowledge from microclimates across the West: Wendy Proud, Karrie Reid, Bracey Tiede, George Hull, Brian Kimble, Richard (RG) Turner, Janet Sluis, Pete Villeux, Randy Baldwin. Thanks especially to Judith Phillips and Scott Calhoun. Thanks also to Hunter Ten Broeck for sharing your time and vast expertise. Thanks to three very special women. Landscape designer Linda Chisari unwittingly started me down this path when she introduced me to the term dry garden when we first met in 1992. My dear friend, author and gardener Jan Smithen, has shared her enormous knowledge over the many years that we have known one another—her great enthusiasm inspired my own. Finally, long ago, when a door closed, Mary Hellman James opened a window and pulled me into her world. I am eternally grateful to her for that and for her many years of friendship and support.

The biggest thank you goes to my family for their patience, love, and support, and for indulging me in this plant passion that has been my driver for so many years. Please know that you are far more important than any plant I've ever come across.

INTRODUCTION

Drought, drought, and more drought—dry weather, of course, is to be expected in most of California, Arizona, Utah, Nevada, and New Mexico, but it is only getting worse. Recent years have brought the worst drought California has ever experienced, and along with it a wide array of mandatory water cutbacks. Arizona and California seem perpetually mired in battles over the water of the Colorado River. That water will diminish as climate change promises hotter temperatures and reduced water supplies across the West.

Drought has been an issue in the Southwest for a long time, but increasingly across North America and around the globe, population growth and global warming are making water more and more a focus of sustainability. And as water becomes more precious, gardens suffer. We need to make significant changes in our aesthetics, our attitudes, our plant choices, and our gardening practices.

This is what I've spoken, taught, and written about for decades. And throughout that time, I've found that gardeners' biggest fear of waterwise gardens is the misconception that these are brown, lifeless, and colorless gardens—but nothing could be further from the truth.

Low-water gardens buzz with life. They are bright, brilliant, colorful gardens with as much interest and variety—and in some ways more—than any other gardens. In fact, color and low water go hand in hand. This is something I've known intuitively for many years. I trace it back to a trip my husband and I took to Santa Fe, New Mexico, long ago. We walked up Canyon Road (a street now infamous for its profusion of artist studios), and as we strolled, I kept noticing the gardens. They were modest, some simple, narrow planting beds tucked up against adobe walls and holding just a few plants: a red- or pink-flowering penstemon perhaps, with a blue-flowering cornflower and a trio of royal purple bearded irises. There weren't many flowers, but they stood out as brightly and distinctly as if there were an entire mass of color. The bright greens, silvers, purples, blues, pinks, and yellows that filled those beds lit up in the desert sun.

In the years since, I've thought often about those tiny, dry gardens and the huge visual impact that resulted from the combination of three factors: the flowers and leaves were deep, intense,

Introduction

saturated colors, while the background earth tones were equally rich, and the sky was a clear, bright, intense blue. Together, the effect was dazzling.

Common Misconceptions About Low-Water Gardens

In my travels throughout the Southwest, I've visited countless color-filled, low-water gardens. It is high time that we set the record straight about what can be achieved, even with increasing water constraints.

Three central myths about waterwise gardening need to be corrected straight away.

Myth #1: Low-water landscapes are brown, lifeless, and colorless

Low-water gardens are anything *but* brown, lifeless, and colorless! In fact, plants from dry regions of the world seem to evolve the most colorful and interesting flowers, the most varied and colorful leaves, and attract an amazing array of wildlife.

My own garden and those I design for clients are filled with riotous color—hot colors—alive with butterflies, lizards, rabbits (though I wish we could get rid of them), and birds, including the ever-present hummingbirds whose iridescent throats glisten garnet and emerald as they dart about the garden.

People are amazed to see the color and variety. Not long ago, I appeared on a local television program to talk about

Low-water landscapes can be colorful, energetic, and exciting. There's nothing brown or lifeless about them.

Introduction

Dense, layered plantings and carefully selected plants create lush gardens.

waterwise gardening. I arrived at the studio with my truck full of plants, and assembled the display to demonstrate several color themes. The public relations person who arranged my appearance had a decent grasp on the concept of a low-water landscape, but wasn't herself a gardener. She was absolutely amazed by the rainbow of colors in my show display.

I'm not the only designer who creates colorful, active, low-water gardens, of course. There is a group of professionals, mostly in the West, who have been leading the way. Their gardens, along with

Dry-growing plants, succulents and nonsucculents alike, with colorful foliage and beautiful blooms, arranged in careful layers against a green background—this typifies the hot colors that can be found in a dry garden.

some wonderful homeowner-designed gardens, are featured in these pages.

Myth #2: Low-water gardens are scrubby and scrappy rather than lush and plant filled

This is just entirely wrong. There are native low-water habitats that, to the uneducated eye, might look scrubby and scrappy in the dry heat, but gardeners can create lush, vibrant gardens by selecting plants carefully, by balancing shades of green, and by massing plants and placing them closely enough to cover the ground, but not so close to require constant pruning to separate them.

"Lush" is when a tall grass emerges from a mounding shrublet, or when a small tree is underplanted with perennials and bulbs. "Lush" is an effect, not a plant type. "Lush" does not require water.

Myth #3: Low-water gardens are all rocks and desert

I often hear gardeners complain, "low water is okay, but I don't want my garden to look like the desert!" This image hearkens back to the so-called Palm Springs– and Las Vegas–style landscapes of the 1950s and 1960s. Those front yards were done up in sharp gravel, laid out in swirls or other geometric patterns, and edged in red scallop brick. The gravel color palette was eye-blinding white with bright teal, or muddy coral, typically accented with a single saguaro cactus or tall yucca. In short, it was yucky.

In a low-water garden, rocks—real rocks, not faux dyed ones—become architecture that balances plants. Large boulders define contours and demark dry streambeds filled with gradations of cobble. These streambeds collect and hold water, so it has time to percolate into the soil where it is banked for access by plant roots. Rocks that appear to emerge from the earth serve as places to sit and rest for a moment, or to separate one area of the garden from another.

A few years ago, I designed a garden for a couple with very different ideas about what they wanted. They agreed on low water, but while the husband wanted a tapestry of succulents and unusual plants, the wife wanted what she called "frou-frou." I put on my marriage counselor hat, and negotiated an eclectic front garden for him, with a "frou-frou" rear garden for her. Both were to be colorful, textural, and low water. Six months later, the wife gushed over how much she loved the front garden (the one I designed for her the husband). She had no idea, she said, that a low-water garden could look like that. And, she insisted I add some of those same plants to her back yard. Now, several years later, both front and back gardens are waterwise and wonderful. They are both pleased.

It's not hard to explode these myths; it takes some really great examples and good information shared with homeowners, professionals, landscape designers, landscape architects, contractors, installers, and maintenance people. That is what I have set out to accomplish with this book.

USING COLOR IN THE GARDEN

Our Mediterranean and southwestern gardens have a common color backbone of dark green, olive green, silver, gray, and blue-green leaves, combined with the soft browns and mahoganies of branches and trunks, all in clear white light under intense blue sky.

The late garden designer Ingrid Rose once told me that every garden she designed included five to seven shades of

In Santa Fe, New Mexico, gardens need not be big to have a huge impact. A few intensely colored blooms and some foliage against an earth tone background is all you need.

green. This practice came from painting classes she took as a young adult in her native England. Upon moving to the San Diego area and translating her artistry from paints to plants, Rose carried the concept of five to seven greens into her garden designs.

By keeping Ingrid's lesson in mind, we too can create dry-climate gardens with color that rivals—or outshines—any temperate or tropical garden in the world.

Designing With Color

Low-water gardens and plants with intense, saturated colors go hand in hand. The more control you exercise over the garden's colors, the more harmonious and intentional the garden looks. Adding negative spaces (areas that de-emphasize color) balances the intense colors. These techniques help avoid randomness and visual chaos in the garden.

Some gardeners and garden designers rely on foliage for color, others use flowers to color the garden. Each has its role, of course. Foliage is forever color, present regardless of the season. Even plants that lose their leaves in fall can go out in a blaze of color. Flowers are ephemeral, but their blooms liven the garden and bring us immense joy when they appear.

In my work, I always use both flowers and foliage. My design process starts with the space planning, soil contours, and hardscape that together create the structure of the garden. Once that is done, I make a large list of potential plants whose flowers and foliage best play off the color and style of the home and its surroundings. And of course I take into account the plant and color preferences of the garden's owners.

Color Is Different Here

Here in the arid West and Southwest, rain is scarce and gardens are dry; to conserve water, leaves tend to be too small, too narrow, and too spare to create the intense green background typical of areas with higher rainfall. So rather than playing off a leafy green background, our gardens play off the bright blue sky, silver, olive, dark green leaves, the clear white light, and the soft brown of the earth.

The best shades for our gardens echo those of our neighbor Mexico, whose "fiesta" color palette includes vivid orange, deep red, lemon yellow, deep purple, cobalt blue, and other strong, intense, fully saturated tones. In Arizona and New Mexico, these tones are often balanced with their more earthy versions as well.

These intense colors are what I think of as "hot colors." Hot colors are not limited to the reds, yellows, and oranges of the color wheel. They include the entire range of strong, deeply saturated colors. Hot colors are popular in other hot climates as well. The colors of the Caribbean, for example, are hot tropicals: bright aqua, cobalt blue, lime green, golden yellow, tangerine, raspberry, and grape. In Morocco and other regions of the Mediterranean, one sees rich jewel tones, tempered by shades of ocher and terra cotta.

I quickly divide my plant list into groups based on how thirsty each is, along with plant preferences for shade, sun, and so on. Often a list gets parsed into two or three lists, each customized for a particular area of the garden. Streetside gardens, for example, are a garden's public "face," so the plants I use there are tougher and bolder than, say, the plants chosen for a courtyard garden. While each area of the garden gets its own distinctive look, there needs to be some continuity. So, while there is an individual plant list for each area, there are some plants that appear on every list.

Start With A Color-Filled Backdrop

Designing a colorful, low-water garden starts with the home. Since the home's exterior walls serve as the garden's backdrop, the colors of each need to work together.

Strong-color gardens really stand out against strong-color homes. White walls work for Moroccan- and Spanish Revival–style architecture, but in most other situations they contribute very little to a dry-climate garden. Always consider painting before you start planting.

My husband and I own a vintage 1970s home in Encinitas, California, just north of San Diego and about three miles from the beach. When we purchased the home years ago, it was like many other houses in our neighborhood: aged, dingy, and boringly white. As we considered stucco colors, that memory of the gardens I saw in Santa Fe flashed through my mind. You can guess what happened next. With some persuasion, my husband agreed to my painting different shades of terra cotta in patches on the house's western, southern, and eastern walls. For weeks we looked at the colors in light from each direction. We looked at them at different times of day. Taking our time and looking at the colors in all those different exposures

A flower the color of this *Sphaeralcea* (globe mallow) would look garish in a temperate-climate garden, but in the clear light and bright blue skies of arid climates it works beautifully.

In my garden, terra cotta walls and a green metal roof are the backdrop for hot-colored plants from every arid corner of the world.

helped us find just the right shade of terra cotta, to which we added accent walls of pale, buttery yellow. We chose a green metal roof and burgundy window frames. The color combination might sound a little odd (at least one neighbor thought it was), but I knew what would happen once I started planting. The colors serve as the perfect backdrop for my very eclectic and very colorful garden.

Our garden is filled with an enormous and ever-changing variety of low-water plants, all of which shine against the deep terra cotta walls. It is so filled with color that people constantly stop me to comment and ask about the plants. Changing the color of our home was a bold move but it proved to be a great decision. It reinforced the idea that one of the keys to creating a colorful garden is starting with a colorful background.

I think of terra cotta as the "blue jeans" of colors when it comes to gardening—it goes with plants of any color. More than that, though, terra cotta enhances other colors. With terra cotta walls, plantings can be minimal, even

Brown tones make great backdrops too. Here, I designed a garden of succulent and broad-leaved plants to shine against café au lait–colored stucco and siding, accented by mahogany-stained wood details in this Leucadia, California, garden.

in Mediterranean-style gardens. Most brownish earth tones have the same effect: browns, deep tans, umber, sepia, cocoa, cinnamon, café au lait, caramel, ochre, adobe, and so on. Greens, especially the sage end of the green spectrum, also enhance garden colors. What color is your home? What garden colors will best complement it? The options are endless.

Color Schemes

Some people prefer just a few colors in their gardens, while others want every color of the rainbow. Different tastes and different perspectives are part of what make gardens and gardening so interesting. Spend some time thinking about your preferred color palette.

Single color To be clear, there are no *true* single-color gardens. Even the historic white garden at Sissinghurst Castle in England is not entirely white. The white flowers are a stark contrast to the red brick walls that surround the garden. Leaves are deep greens and silvers. Flowers have red bracts or yellow centers, with pale blue leaves. Without those

At Loseley Park, the white garden is enhanced by varying shades of green mixed with burgundy foliage and yellow accents.

FAR LEFT: Clear white roses with their bright green foliage are a beautiful foil for old red brick walls in the famous white garden at England's Sissinghurst Castle.

LEFT: At Sissinghurst Castle, the ghostly white *Eryngium* flowers steal the show.

San Diego designer Dinah Grisdale's yellow-theme garden includes apricot, greens, pinks, and reds to enhance the yellow tones.

color contrasts, the white flowers would look very plain. And while the white garden at Sissinghurst is better known, I prefer the white garden at Loseley Park in Surrey, England. It is much more interesting, in part because it is planted with more color contrasts.

While white gardens are beautiful in England's moodier climate, they seldom work in Mediterranean- or desert-climate gardens. In the bright sunlight, the whites are too glaring and too bland. Instead, try a yellow-theme garden. That's what San Diego designer Dinah Grisdale created for her front garden a number of years ago. For her space, she chose *Aeonium* 'Kiwi', with its yellow, green, and pink foliage, a green and yellow variegated aloe, yellow-flowering *Phlomis fruticosa* (Jerusalem sage), chartreuse-flowering *Euphorbia rigida* (gopher spurge), and yellow-blooming nasturtiums. To set off the yellows, she added apricot-flowering nasturtiums, pink-flowering pelargoniums, and sunset-colored succulent *Kalanchoe luciae* (flapjack plant). The contrasting colors enhance the yellows.

For this walled garden, designer Laura Morton chose a cabernet wall pigment matched with green- to red-leaved *Vitis* 'Roger's Red' (California grape) and *Cercis canadensis* 'Forest Pansy' (eastern redbud), which is waterwise along California's coast. She underplanted the tree with a cerise-flowering geranium—all variations of the same color.

Some desert gardeners bemoan the prevalence of too many yellow flowers in the plant palette. Instead of planting a yellow-theme garden, look for opportunities to create other tone-on-tone combinations like this *Chilopsis linearis* (desert willow) blooming against an earthy lavender-colored wall in Tucson, Arizona.

Using Color in the Garden

Designer Scott Calhoun's office garden in Tucson features orange walls that contrast with green leaves, intense blue sage flowers, and the feathery blue-green foliage from *Yucca rostrata*.

Tone-on-tone Desert gardeners sometimes bemoan there being too much yellow in the native plant palette, but yellow is not the only option. Color-theme gardens can be purples, reds, or just greens. Some of the most interesting color gardens are tone-on-tone, where foliage, flowers, garden walls, and accessories are all different shades of the same color. Los Angeles–area garden designers Laura Morton and Judy M. Horton each create exquisite gardens using this technique. It can take some practice to do well, but their examples can be followed to great effect.

In Judy Horton's case, she chose plants to match her home's existing wall colors. Laura Morton selected both the wall color and the plant color palette to work together. All of their gardens feature plants with intense colors.

If you find a flower or foliage whose color inspires you, take it to a paint store. They have the technology to match paint color to the color of any leaf, flower, or stem—or any other object for that matter.

ABOVE LEFT: This contrasting color combination features red pomegranate (*Punica*) blooms spilling over a soft green wall.

ABOVE RIGHT: This red, orange, and yellow color combination includes pumpkin-colored *Arctotis* The Ravers 'Pumpkin Pie' (African daisy) in the foreground. The emerging flower stalk of the variegated yucca has a reddish tone that echoes a background of red- and orange-flowering *Lobelia laxiflora* (Mexican cardinal flower).

LEFT: *Parkinsonia* 'Desert Museum' in bloom with *Salvia chamaedryoides* (germander sage) in my front garden in Encinitas.

A coral-colored wall sets the tone for this hot combination of orange-flowering *Aloe striata* against a golden mound of *Sedum nussbaumerianum*. Notice how the round, red blades of *Kalanchoe luciae* in the foreground enhance the gold and orange tones.

Using Color in the Garden

Designer Amelia Lima created this spectacular arrangement of succulents with broad-leaved plants. Foliage and ceramics in cool blues and greens are highlighted with a bit of yellow and chartreuse.

San Diego–area landscape designer Scott Spencer planted this cool-color combination that relies on green-, pink-, and blue-toned succulent leaves with purple flowers from nonsucculent fall-blooming *Salvia leucophylla* (Mexican bush sage).

Red cactus spines, blue agave blades, and the flowers of *Encelia farinosa* (brittlebush) create a primary color scheme in this desert garden bed.

Contrasting colors create great excitement in the garden. These are the colors that are opposite each other on the color wheel and whose effect is enhanced when they are used together. Orange and blue, yellow and violet, and red and green are classic pairs of contrasting colors. The more saturated they are, the more vibrant in our hot, direct light (and the more important it is to leave some negative space in the visual field). Notice how your eye is drawn to where contrasting colors connect and overlap. These "color intersections" are part of what makes color-filled gardens so exciting.

Warm colors (red, yellow, and orange) are especially well suited to our bright skies and clear light. These colors

Using Color in the Garden

If you can put every color into a garden, and it works, why not do it? That's what designer and landscape contractor Gabriel Frank did for this garden on California's central coast.

look gaudy in temperate climates where skies are more grayish and overcast. But in the bright light of an arid garden, they are just right.

Cool colors When I was a child, all the girls I knew decorated their bedrooms in the traditional pink and baby blue. I always liked green with purple. Green and purple, along with blue and all the shades in between, are traditionally considered "cool colors."

A hot-color garden can include these cool colors too, as long as they are highly pigmented and deeply saturated. Add a touch of yellow, orange, or another contrasting color to make those cool colors pop.

Primary colors You might think a combination of primary colors like red, yellow, and blue would be a bit dull and ordinary, but they too look magical in hot-color gardens.

I have come to love the combination of pink with orange, though some people find it jarring. The flowers (technically bracts) on this variegated *Bougainvillea* 'Bengal Orange' open orange and then turn pink. For balance, I added yellow-flowering *Calylophus drummondianus* and blue-bladed succulent *Senecio mandraliscae*.

Despite my initial resistance to pink in my home garden, I eventually found a successful combination. Here, the pink African daisies (*Arctotis* The Ravers 'Pink Sugar'), work beautifully in combination with the blue-leaved and burgundy-spired *Melianthus major* (honeybush), orange-blooming *Kniphofia thomsonii*, and chartreuse-flowering spurge (*Euphorbia*).

Using Color in the Garden

The dusky gray, mauve, and rose foliage of *Adenanthos cuneatus* 'Coral Drift' combined with the blue-gray foliage and blue flowers of *Dianella caerulea* 'Cassa Blue' pick up the gray tones in the trim on this home in Encinitas.

The bright yellow globe flowers of *Craspedia globosa* (Billy buttons) offer a stark contrast to the dusky silver fronds of *Melianthus major* 'Purple Haze' and *Cordyline* 'Festival Grass'.

One, two, three, or every color
Why stop? All colors are welcome in your garden. Keep an open mind and be willing to try colors, even those you might otherwise resist. When I started my garden, I adamantly rejected any plant that bloomed pink. That stance lasted just a few years, until I discovered the winning combination of intense, saturated pinks with plants that have blue leaves, along with chartreuse and burgundy flowers.

Unusual color combinations Once again, it is important to keep an open mind. Orange and pink are quite close on the color wheel, and their similarity makes them notoriously difficult to combine in the garden. A few years ago, however, I set a challenge for myself to find a way to make them work together.

The owner of this beach cottage inherited a set of outdoor furniture in an unusual color combination. Rather than discard or paint the furniture, garden designer Debra Carl used the colors as the palette for a new garden.

Eureka! Once I added bits of yellow, purple, and blue, the oranges and pinks combined together beautifully.

I recently designed a garden for a home painted a very pale butter yellow with charcoal gray trim. I assembled a plant color palette of deep green, silver green, blue green, bright green, cream, pink, lavender, and burgundy. I then added in butter yellow and melon orange for a color pop. I intentionally avoided sulfur yellow and scarlet reds, as they would have been too jarring to look at.

Strong contrasts Using contrasting shades and intensities is another way to add color and interest to the dry garden. Paler colors against a saturated background and strong colors against a pale background both create drama and interest. This is an opportunity to incorporate colored foliage—red, silver, golden, and variegated. These are the colors that remain constant year-round in mild-climate gardens.

Designer Gabriel Frank created a garden of strong contrasts—in both color and structure—for clients in Morrow Bay, California. Frank makes excellent use of red variegated conebush (*Leucadendron*) foliage as a background for black-colored aeonium rosettes, blue-gray *Euphorbia rigida*, and other brightly colored succulent and nonsucculent foliage.

Dustin Gimbel created this colorful presentation using black *Euphorbia* 'Blackbird' (top) as a backdrop for the pink, teal, silver, and green foliage of succulent *Graptoveria* 'Fred Ives'.

ABOVE LEFT: Colored walls can change the ambience of a garden. In this Albuquerque, New Mexico, garden, a pink stucco wall and sky blue fence are the background for a collection of found objects—rust, brick, and stone.

ABOVE RIGHT: Alan Richards color matches walls, bench, and pillows to the blooms and foliage in his Tucson-area garden.

LEFT: Red trim and cobalt blue pots brighten the sage green facade of Carol Anne McDougal's Tucson garden.

The deep mahogany tones of rusted metal create a rich, colored background for pale stones and silvery and blue-green foliage in this Phoenix garden.

Color Architecture

As mentioned earlier, plants are only one way to bring color to your garden. Bright colors can also come from other garden elements, too.

Walls of contrasting colors have been growing in popularity—and rightfully so. Colored architecture in the garden makes a big statement. Big blocks of color open up a world of possibilities for designing garden spaces.

Stucco, wooden fences, and rusted metal panels are a few options for creating in-garden backgrounds for plantings. Alternatively, these vertical surfaces can serve as focal points, especially when embellished with found objects, artwork, and other interesting elements.

Pottery

Pottery is a beautiful way to bring color to the garden. Colorful plants and colored glazes all contribute their brilliance. A good way to start a container collection is by focusing on one color family—red or blue, or ochre, for example—chosen to fit your garden's color scheme or to contrast with it. Another option is to match or contrast the color of your pottery to the color of your home's trim.

Sticking with one color family creates unity. Varying the size and shape of pottery creates interest. So shop for a collection of containers, three or more, in different sizes, different shapes, even different surfaces—but all in the same color family.

In this stunning collection, designer Judy Horton sticks to terra cotta pots, each filled with succulent *Aeonium* 'Velour'.

In this collection of multicolor vintage pots, the contrasting blue, green, and yellow play off a single, shell-covered container in similar tones. The aqua bench adds to the color scheme, while the smooth-surfaced pots and bench contrast with the rough shells. Notice that pots are all the same scale and similar shape as well. Plants, in this case, hardly matter.

Using Color in the Garden

Los Angeles landscape designer Nick Dean met the challenge of a blue on blue glaze pot with the similarly colored succulent *Echeveria* 'Perle von Nürnberg' beneath the spectacular burgundy rosettes of *Aeonium* 'Zwartkop'.

Los Angeles designer Cheryl K. Lerner mixes different-colored vintage pottery in her own garden to create a cohesive collection.

The exception to this rule of thumb is terra cotta (again, terra cotta shows its value as the "blue jeans" color). Terra cotta pottery mixes with any color ceramic glaze. Its matte surface also adds contrast to the high gloss of most glazes.

When it comes to planting pots, the principles are the same. Choose plants whose leaf and flower color are in the same color family as the pot, or select those that create a strong contrast with the color of the container.

Not every pot needs to be planted. Bright-colored, unplanted pots can add just the right touch of color and architecture to a garden bed.

If you like the look of just one spectacular plant per pot, take care selecting the size and shape of the pot. Balance taller pots with taller plants. Shrubs are great in pots and most shrubs will survive in a pot for at least a few years. Wide, low pots are good for plants that are more diminutive, have shallow roots, or want to cascade over the sides. Combine those single-plant pots so you have a big pot with a big plant surrounded by several medium-sized pots with medium-sized plants and a number of smaller pots with plants matched to their sizes as well.

Another approach is to plant larger pots with combinations of tall, wide, and cascading plants. Together, these plants form a sort of miniature garden where color and structure, once again, are critical considerations.

In the "I love every color" category, find unity by creating a theme collection based on something *other* than color: a

The colors of designer Cheryl K. Lerner's Mediterranean-style courtyard come from vintage furniture, ochre-colored walls, ceramic pots, and a magical assortment of colored accessories. Plants are secondary or even tertiary.

Using Color in the Garden

collection of historic Bauer pottery, for example, or containers that are all similarly shaped.

Must all pots be planted? Definitely not. Try placing a large, empty brightly glazed container in a garden to serve as garden architecture.

Details

There are still more ways to add color to dry gardens. Some gardeners make use of colorful furniture to create spots of color in the garden. Wood is the standby for garden furniture. It is easy to paint and repaint, easy to move, and looks absolutely natural in the garden, regardless of the garden style.

Metal furniture, whether modern or vintage can be painted nearly any color. Powder coating is another way to color metal furniture and it offers more rust resistance, which is especially important along the coast.

Weather-resistant resin wicker furniture seems to come only in passive neutral tones. Dress it up with paints made specifically for plastic. And on the occasion that you find colored wicker or resin outdoor furniture, buy it!

What else can you pull into the garden to add color? Candles in winter (they might melt in summer), colored bottles, colored stone, pieces of art, and so on. Let your imagination run wild!

OPPOSITE TOP LEFT: Reds, soft greens, and teals are featured in this corner vignette that showcases a vintage metal chair, watering can, *Agave attenuata*, and teal-colored glass mulch.

OPPOSITE TOP RIGHT: The corner of a hot, covered desert patio gets most of its color from a pair of vintage wooden chairs whose slats are painted in sherbet hues.

OPPOSITE BOTTOM: Turquoise slag glass, ceramic orbs, and an old glass telephone line insulator match the color of succulent *Agave parryi*.

ABOVE RIGHT: Yellow glass orbs in a small pond heat up the colors in this dry garden in Long Beach, California.

RIGHT: A simple collection of small potted plants and vintage bottles in the same color tones brighten an outdoor kitchen in the California desert.

DESIGNING FOR PLANT STRUCTURE AND TEXTURE

Garden designers and landscape architects talk about gardens in terms of hardscape and softscape. Hardscape is a catchall for walkways, decks, stonewalls, pools, and other built elements of the garden. Softscape refers to the plants.

Successful gardens play hardscape off softscape and vice versa. Each garden featured in these pages was chosen for its colors *and* for its balance of hardscape and softscape. It is important to guard against using too much hardscape. Too much can feel harsh and hot; however, too little can lack structure and cohesion. There is a world of literature to help you strike the right balance between hardscape and softscape, and much of it focuses on the texture and structure of plants. In dry gardens, plant structure and texture help support color, too.

Plants best adapted to dry-climate gardens have a very specific set of characteristics: some are fine leaved, others broad, some succulent, some green, others silver, some burgundy, variegated, and so on (see the Dry Gardening How-To chapter to learn how leaf architecture and texture relate to water conservation).

Combining plants based on their texture and structure takes a practiced eye, and done well, the effect is stunning. Done poorly, the garden looks like a tangled mess. Keep in mind that gardens are layered affairs. The goal is to emulate nature by including tall, wide, upright, narrow, and low plants. Without one of those components, a garden can look unfinished.

Many gardeners are afraid of large plants, and are fixated on groundcovers. But nature doesn't cover a habitat in just groundcover, and neither should we. The best gardens feature a full complement of plant sizes, from an overhead canopy to a layered understory, whether dense or sparse. In the words of a nursery friend of mine, "all plants cover the ground!"

Every garden needs a tree, or two or three or more, depending on how large the space is. Even a courtyard can support a tall shrub trained into tree form. Beneath the tree are the larger shrubs that form the backbone of the garden. Flowering shrubs are my first choice, but flowers are not always necessary.

And unless you live in an ultramodern home, please *don't* clip or shape your shrubs. That isn't how they grow naturally and that isn't how they look their best. Space shrubs, and all plants, based on their mature dimensions, then let them grow to those sizes, naturally.

This Bernard Trainor–designed garden in Palo Alto, California, complements an angular, modern-style home. Solid, garden architecture supports Trainor's design. He softened the angles by using broad, soft surfaces like this decomposed granite bocce ball court. Plantings are lush but low water, and include California natives along with plants native to South Africa and Australia. Their textures are a balance of fine-leaved, broad-leaved, and succulent-leaved.

Designing for Plant Structure and Texture

Rather than planting a hedge, San Diego designer Amelia Lima layered succulent and nonsucculent plants to create a colorful and highly textured privacy screen for her streetside garden.

As you choose broad-leaved plants (the "nonsucculents"), pay attention to plants that have small or fine-textured leaves. Small leaves and fine leaves are common adaptations for arid-climate plants. Since leaves lose moisture through the surface of their leaves, reducing that surface is a way to conserve water, and survive heat and drought.

When it comes to combining plants for a garden, too many plants with small or lacy leaves can be hard to look at. To balance those plants, include plants with wider, even leathery leaves, along with succulents.

Many gardeners want one of each type of plant. While I understand (and share) the impulse, a "one of this and one of that" approach to gardening never

Designing for Plant Structure and Texture

Santa Rosa designer Michelle Bellefeuille takes a painterly approach to garden design. In the tiny front yard of this pink home with teal trim, she focuses on color, form, and texture. She mass planted *Festuca idahoensis* with blue-leaved *Epilobium* 'Schieffelin's Choice' that blooms coral red in spring. If she had used one of each plant or mixed them together, the effect would have been chaos. Instead, she grouped them to create a beautiful, textural contrast.

works out well. Instead, mass plants in groups of threes or even fives, then repeat groupings to create a naturalistic and cohesive look. Three groupings of three plants each is a good starting strategy, especially for smaller plants.

Within that repetition, there is room for variation. Specimen plants—those few special, large, sculptural, colorful, or otherwise notable plants—get special treatment. The massed plantings create the framework while the single specimen serves as a focal point.

It's fun to experiment with texture, even while keeping the color constant and varying the shade. Low-water grasses and succulents make a wonderful textural combination when paired together. Mix plants of different heights and structures too—mounding, upright, spreading, broad, and so on. Place mounding perennials in the foreground, taller perennials in the center juxtaposed with broad agave blades or round *Echeveria* rosettes, all against a background of evergreen shrubs. Try combining bright green with ice green, burgundy with cabernet, silver with blue-gray. Add a splash of contrasting yellow or chartreuse to enhance the effect.

Grasses and grasslike plants add an upright component to the garden and their slender blades undulate in the breeze, bringing movement to the garden. Movement adds another dimension that speaks to our senses. It is the perfect complement to visual structure and texture.

In smaller gardens, pots can create structure and texture, too. Play off their architecture by adding wispy plants to smooth-glazed pots, or smooth-leaved succulents to rough-textured pots. Plant your tallest or largest pots to focus attention on special plants.

As you look for opportunities to play with texture and structure, also consider your garden's "borrowed view"—the view that surrounds your garden. Are there trees? Distant mountains? Gorgeous vistas? Is your neighbor's beautiful garden visible from your own? Choose colors, textures, and plants for your garden to reflect and incorporate those elements. You'll find that your garden seems like a part of the larger space. This is a classic approach to expand the garden's look and feel. It also helps integrate the garden to its site.

Designer Nick Wilkinson is a plant artist. This Santa Barbara hillside garden had an existing canopy of deep green–leaved oak trees. Beneath the trees, he added a single tall tree aloe as a focal point, surrounded by a mass of pink, gold, and green succulent *Euphorbia tirucalli* 'Sticks on Fire'. Red-flowering *Anigozanthos* 'Big Red' (kangaroo paw) grows in front of the euphorbia, followed by the blue-green leaves of succulent *Aloe plicatilis*, and finally a yellow variegated *Yucca* 'Bright Star'.

In this garden bed, mounding forms are repeated to create cohesion, including low-growing succulent aloes and agaves with a mass of grasslike *Carex testacea*, and gold-and-red *Gaillardia* (blanket flower), a dry-growing perennial that blooms from spring through the heat of summer.

In this tone-on-tone vignette, the focus is on contrasting textures from the smooth-surfaced, succulent rosette *Echeveria* nested in a background of tiny, round, fuzzy-leaved ornamental *Origanum dictamnus* (dittany of Crete). Notice how the creeping tendrils of succulent blue chalk fingers (*Senecio mandraliscae*) add some geometry and enhance the textural effect.

ABOVE: In his personal garden, designer Dustin Gimbel painted a detached garage dusky purple, then covered it with the round-leaved, vining Aristolochia gigantea. At the base of the wall, he composed a bed of blue-fronded Melianthus major 'Purple Haze' and Coleonema pulchellum 'Sunset Gold'. A specimen Euphorbia cotinifolia features round, burgundy leaves and will eventually be as tall as the garage. Opposite that, Gimbel placed a green and cream variegated ponytail palm (Beaucarnea) whose straight, upright form balances the composition's colors and textures.

LEFT: Upright, grasslike Kniphofia (red hot poker) in brilliant orange contrasts with the cascading, round blue flowers and rounded leaves of Convolvulus mauritanicus (ground morning glory). Masses of grasses and repeated red hot poker in the background enhance the feathery texture of the design.

OPPOSITE: Make use of what the borrowed view has to offer. Nick Wilkinson pulled the borrowed view of dense, chaparral-covered hillsides into this garden by planting just a few specimens of Euphorbia ammak var. variegata that make a statement with their thick, pale green, upright succulent stems. Grasses and other grasslike plants sway in the breeze, bringing movement to the scene.

Remember to leave some negative space in your garden, as well, to give your eyes a place to rest. Lawns once served this purpose, but thirsty lawns simply don't belong in our water-challenged gardens. Lawns are best relegated to parks and other common-use spaces. There are plenty of other opportunities for negative space, however. Treat walkways as negative space. Allow room for mulch—rock or organic—between plants. Add a dry streambed. Try not to fill *every* corner of the garden with plants, even though it can be challenging to resist this urge.

While gardens in dry Mediterranean climates tend to be intensely planted, the same design concepts apply to the more sparsely planted dry-desert gardens. Color, structure, texture, and negative space all need to work together to create a cohesive whole.

The view from my front door shows a lush planting with layers of trees, flowering shrubs, agaves, grasses, and grasslike sedges (Carex). On the right, a vase-shaped ochre-colored pot is planted with a tall, red-leaved cordyline that developed multiple heads over time. I incorporated the borrowed view by using the neighbor's garden across the street as the background for this vignette. The line between the two spaces is completely erased.

In the drama of this spare desert planting, layers of plants in different shapes, textures, colors, and sizes all work together and play off the sage green walls and red trim. Notice the many shades of green, some accented with tiny yellow flowers, others with purple blush blades or with blond spent foliage. Here, one garden blends into the next, and into the surrounding habitat. This is the essence of a brilliant naturalistic landscape.

GARDEN GALLERY

The concepts behind using color in the dry garden work with gardens of every style and every size, from urban condominium to ranch style rural, modernist cube to earthy adobe, and everything in between. The gardens featured here were chosen to represent that enormous variety and show how, despite the fact that they are spread across a large geographical area, they adhere to the same basic principles.

Every garden description explains how the designers thought about using color in the garden, as well as their overall design intent, along with their water-saving strategies for the space. Each is also accompanied by Facts and Figures, an at-a-glance summary of the garden's physical conditions. When I visit a new garden, I use that information to "read" the garden, to better understand its design and plant palette. This information provides a context intended to help you translate the lessons of these gardens to your own. They tell you about each garden's climate (including summer highs and winter lows), how much rainfall it receives, how humid the environment is, the soil type, how well the soil drains, and how the plants are irrigated, and so on.

Pay special attention to when each garden gets its major rainfall. If it is a summer-rain garden, for example, the plants in that garden are evolved with summer rainfall. If you live in an area that is dry all summer, you'll need to water some of those plants, especially the smaller and nonsucculent plants, through summer to keep them alive. This is not necessarily a negative, but knowing this information helps you anticipate the plants that will or won't thrive in your garden. Also notice how humid the garden is and what time of year it is most humid. When the air is drier, plants lose more water to the atmosphere and require more irrigation to replace that lost water.

At the end of each garden's description, you will find a short list of key plants for color in that garden. Use the Facts and Figures information, along with the Plant Directory, to assess how well each plant will work in your garden.

In the end, however, trial and error is a huge part of the gardening process, so if you like a plant go ahead and try it, even if it seems to come from a totally different kind of climate. The thrill of success always overshadows the agony of defeat.

Garden Gallery

Knock Your Colored Socks Off

Ring the entry bell to Patrick Anderson and Les Olsen's Fallbrook, California, garden, and the tall, redheaded Anderson is likely to greet you wearing teal-colored trousers and a melon orange shirt. His colorful persona, both inside and out, is only a hint of what's to come.

Anderson's tour starts at the streetside garden he designed a few years ago. The gated driveway is flanked with angular planter beds stuccoed dusky sage green. The structure and geometry play off mass plantings with a limited plant palette. Still, each bed is filled with a dramatic combination of succulents and cacti. Yes, cacti. Most gardeners are afraid of cacti, but when it comes to plants, Anderson fears nothing.

In one bed is a stand of *Pilosocereus pachycladus* (blue columnar cactus) that range from 8 to 10 feet tall. The ridges of each column are lined in what looks like unruly terrier hair. Here and there are enormous grape purple buds that open to papery bone white flowers. A mass of bright orange *Sedum nussbaumerianum*

The gated entry to Patrick Anderson and Les Olsen's Southern California garden is flanked with angular planter beds whose structure and geometry play off dramatic plantings of succulents and cacti in tones of sky blue, bright orange, rusty red, and deep green. Shadow and texture enhance the dramatic plantings.

(coppertone stonecrop) blankets the ground at their feet. Nearby is a colony of large, round *Echinocactus grusonii* (golden barrel cactus), whose spines glow gold in the early morning light. *Aloe* 'Jacob's Ladder' (Dawe's aloe) blooms a bright orange above succulent brick red leaves. A spray of narrow green blades from *Dasylirion longissimum* (Mexican grass tree) contrast the other plants' solid shapes and forms.

From the street, Anderson leads the way through the gate and past the Dawe's aloe and smooth, icy green *Agave guiengola* set into a soft sea of gray-leaved *Arctotis* (African daisies) blooming pale pink and pale orange. Unlike the structured and geometric streetside garden, the inside gardens are flowing and informal, yet they also feature an eclectic collection of plants.

A long driveway that leads toward the main house is lined in massive California pepper trees (*Schinus molle*) that cast a dappled shade over specimen bromeliads growing in oxblood red and cobalt blue pots. All are surrounded in a tapestry of groundcover aloes, more African daisies, and terrestrial bromeliads.

The driveway leads to the heart of the 2-acre property where Anderson and Olsen's once plain Jane Mediterranean-style home is painted bright gold. The home is surrounded by Anderson's collection of exotic flowering trees and shrubs, whose colors complement the gold. This is the oasis area, watered just a bit more so it packs a visual punch. Plantings include narrow "leafless" *Strelitzia juncea* (bird of paradise) that blooms golden orange, as well as a towering 8-foot-tall *Crinum* with dark cherry red blades that measure at least 5 inches across. Beneath a fringed-leaved

Pilosocereus pachycladus is one of nature's true blue plants. Native to Brazil's dry forest, it performs well in frost-free gardens. Its blue, white, and purple color combination is surpassed only by its fabulous architecture, contrasted here against a sage green background and orange-red accents.

Tangerine orange, dusky sage green, deep green, golden yellow, sky blue, and red: these are the colors that welcome visitors to Anderson and Olsen's garden.

The colors from outside flow through the entry gate; here, blooming orange *Aloe* 'Jacob's Ladder' and icy green *Agave guiengola* are set into a sea of African daisies that bloom pink and orange.

The once ordinary Mediterranean-style home is painted bright gold to offset a collection of exotic flowering trees and shrubs. The oasis zone, which is nearest the house, receives just a bit more irrigation to ensure it packs a colorful punch.

Rock and stone pathways meander through this garden, beckoning visitors to explore the combination of bright-colored ceramics, glass, statuary, leaves, and flowers.

Garden Gallery

The garden's color palette includes soft greens, silvers, coral, red, yellow, chartreuse, and orange.

Calliandra surinamensis (pink powder puff) grow succulent *Aeonium* rosettes in green, green and cream striped ('Sunburst'), and bronze ('Zwartkop'). There is tight-clumping *Senecio serpens* (blue chalksticks), chartreuse-flowering *Euphorbia rigida* (gopher spurge), and purple-leaved *Tradescantia pallida* 'Purple Heart' (spiderwort).

As colorful and textural as this plant combination may be, it's not what draws visitors from far and wide. Credit for that goes to Anderson and Olsen's succulent garden sited just across the main driveway. When they purchased the property in 1988, this half-acre slope was a lime orchard. Shortly after moving in, they hired a bulldozer to scrape the tress away. Months of contouring, hauling rocks, making gravel paths, and building wooden bridges followed. When that work was done, Anderson started planting.

For years, Anderson volunteered at the renowned desert garden at the Huntington Library, Art Collections, and Botanical Gardens in San Marino, California. There, he got a first-class education on rare and unusual succulents, along with other

Colored walls draw the eye to both near and distant views. They serve as a background contrast for the garden's foliage and flowers.

OPPOSITE TOP: When a rusted metal sculpture needed a contrasting background, Anderson had a wall built, then painted it turquoise. Bright orange ceramics along with bright orange blooming aloes add more color to this vignette.

OPPOSITE BOTTOM: A stay at Great Dixter, the famous garden of Englishman Christopher Lloyd, inspires Anderson's use of color. He saw the fearless way Lloyd used color, then translated Lloyd's approach to cacti and succulents.

arid-climate plants. As he says, "When I had the opportunity to have a garden, those were the plants I wanted to use. It was the inspiration of the Huntington, but also a practical decision to honor the climate where we live." The garden's succulents are augmented with other desert, Mediterranean, and Australian plants, all of which thrive in 12 inches of annual rainfall and the occasional summer hand watering.

The garden's color palette includes soft greens, silvers, coral, red, yellow, chartreuse, and orange. "The plants dictate the garden's colors," Anderson says, "99% of aloes bloom orange so that is the starting point…. I like succulents because the foliage colors are so interesting—blues turquoise, orange, grays any shade of green, pink. Then you introduce other elements like backgrounds and walls that you can have fun playing off of."

Anderson's ability to play off walls is evident throughout the garden. One of his early efforts is a cobalt blue wall as a deep background against which he planted succulents whose blades and blooms stand out: bright coral orange and yellow *Euphorbia tirucalli* 'Sticks on Fire' (red pencil tree), *Hesperaloe* that blooms creamy yellow, spiny *Aloe marlothii* (mountain aloe), whose coral-colored candelabras open in winter, lower-growing *Aloe* ×*spinosissima* (spider aloe), with red-orange blooms, and bright green aeoniums that bloom yellow in summer.

At the slope's highest point, the men built a summerhouse and painted it the same deep gold as the main house. Inside is Anderson's collection of California pottery, most turquoise or orange. Outside are aloes and agaves. One of the most unusual and striking plants near the summerhouse is *Eucalyptus macrocarpa* (mottlecah), which is shrubby with large, ghostly white leaves. The white is actually a thick layer of wax. Touch it, and the wax sticks to your fingers, revealing the leaf surface as dusky green. Its buds are large, woody capsules shaped like spinning tops, but white and about 4 inches across. As the buds mature, the capsules split and the covers fall off, revealing a brush of deep red fringes with yellow tips.

Turquoise and orange (often in the form of rusted metal) are repeated throughout the garden. A walkway near the bottom of the garden passes by a tall treelike sculpture of rusted metal perched in a bed of turquoise slag and ceramic turquoise orbs. Recently, they added another piece of rusted metal, this one made by Portland sculptor Rory Leonard. The sculpture needed a high-contrast background, so Anderson designed a new wall he painted turquoise, then mounted the piece.

The wall also serves as the backdrop for a sophisticated arrangement of succulents, a gold-and-orange-tone boulder, and bright orange pottery. "The rock was already there," Anderson says, "it was one of the first things to go in long ago, when garden was first done. Once the wall and sculpture were there," he continues, "I needed a pop of color, so I started amassing orange pots…. I had most of those pots sitting around."

The late English gardener Christopher Lloyd helped inspire Anderson's love of color. Anderson stayed at Lloyd's Great Dixter many years ago. In the gardens, he saw many examples of Lloyd's fearless use of color. As Anderson translated those observations to cacti and succulents, he learned to play bluish leaved agaves, for example, off reddish leaved aloes. "I think about complementary versus contrasting colors," Anderson explains. "There's a place for both."

These days, Anderson watches the garden "make itself." The purple-flowering *Felicia fruticosa* (shrub aster), planted years ago, reseeds among the larger aloes and agaves, as does the gopher spurge and the velvet-leaved *Abutilon palmeri* (Indian mallow). "I like the plants to decide where they want to be," Anderson explains. "It makes the garden more natural. I even have aloes planting themselves and hybrids appearing. It's awesome when that happens."

FACTS AND FIGURES

ELEVATION 631 feet
ANNUAL AVERAGE RAINFALL 12 inches, falling mostly between October and March; dry in summer
ANNUAL RAINY DAYS 21
SUMMER HIGH 95.9°F, typically in August
WINTER LOW 40°F in December to January
HUMIDITY Peaks at 85% in August to September, dips to 71% in October
IRRIGATION Most of the garden is hand watered, though overhead sprinklers line the driveway
SOIL TYPE Decomposing sandstone or silty loam, depending on the area
USDA ZONE 10a
DOMINANT PLANT TYPE Succulents mixed with dry-growing Mediterranean plants

KEY PLANTS FOR COLOR

Aeonium 'Sunburst' and 'Zwartkop'

Agave 'Joe Hoak'

Aloe cameronii

Aloe 'David Verity'

Aloe rubroviolacea

Echeveria 'Afterglow'

Echeveria elegans

Echinocactus grusonii

Euphorbia tirucalli 'Sticks on Fire'

Pilosocereus pachycladus

Sedum nussbaumerianum

Yucca 'Bright Star'

Garden Gallery

Pinkie and Blue Boy

In the college area of downtown Santa Rosa, California, there is an intersection where color reigns. Two 1930s-era bungalows face each other diagonally across the street. One is storybook style painted coral pink, the other is ranch inspired with blue wood siding. Both belong to Marcia Coleman and Phil Grinton, who live in the blue ranch. Both houses have gardens designed by their neighbor Michelle Bellefeuille.

Coleman and Grinton hired Bellefeuille to design their front garden in 2011. Coleman, who is so in love with color that she wears rainbow stripes in her hair, made her preferences clear: "I want flowers and I want color," she said. " Color lights me up. It's absolutely essential in my life, and it makes other people happy too."

Bellefeuille is a trained sculptor, printmaker, and plein air painter whose art draws from the natural landscape. Her approach to garden design is similar to her fine arts approach: "I look at masses, shapes, how things fit together," she says. "I start with that as a foundation, then I start looking at plants and working with color."

Low maintenance was also high on Coleman and Grinton's wish list, especially after the couple purchased and restored the pink storybook bungalow to use as a rental.

Santa Rosa is located north of San Francisco and receives about 30 inches of precipitation each year. While that is much more precipitation than Southern California or the deserts of the Southwest receive, it is still a Mediterranean climate

Homeowner Marcia Coleman asked for a garden with flowers and color. "Color lights me up," she says.

Garden Gallery

because all the rain falls between the end of October and early April. For the rest of the year, the region is dry and warm, with summer temperatures peaking in the upper 90s. Wise water management is a must.

Both gardens are small, and both are irrigated with in-line drip irrigation, in this case ¼-inch-diameter lines that snake around individual plants. A thick layer of Sequoia woodland mulch tops the gardens, covering the irrigation and insulating moisture in the soil while protecting roots from the summer heat.

Today, the two gardens are much like siblings, each with strong forms and functions but with its own complement of colors and plant palette. And like siblings, one is exuberant and outspoken while the other is subdued and sophisticated. Yet the gardens share a similarity that springs from their shared maternity.

Blue Ranch Bungalow

A year of sheet composting killed the Bermuda grass that covered this 1,800 square foot garden. The process gave Bellefeuille time to come up with a design that responded to Coleman's request for "color and low maintenance with as many flowers as you can find." Saturated red and yellow were Coleman's colors of choice. By playing off the home's blue façade, Bellefeuille created a palette of primary colors.

The new garden overflows with plants but doesn't feel crowded. The bungalow's blue walls provide color in the vertical plane. Bellefeuille pulled the blue into the horizontal plane with *Ceanothus* 'Centennial' (California lilac), a native groundcover that blooms deep blue in spring, along with blue-flowering rosemary and mounds of brilliant blue-flowering perennial *Lithodora diffusa*. These plants, combined with other

FAR LEFT: The garden features several kinds of roses, some are newly planted and some, like this unnamed variety, are leftovers from the garden's past, yet continue to thrive in the garden's new, low-water diet.

LEFT: The bright yellow blooms of *Coreopsis auriculata* 'Nana' are key to the garden's primary color spring palette.

OPPOSITE: The red and white two-tone flowers of drought-tolerant *Salvia microphylla* 'Hot Lips' anchor the wild riot of color in this garden.

California and Mediterranean shrubs, form a background of small, deep green leaves.

Red and amber carpet roses bloom from spring through fall, featuring masses of fragrant red flowers and amber blooms that fade to peach. Since these smallish rose plants—only 3 feet tall and wide—are established, they require surprisingly little water.

More red and yellow come from flowers and foliage. There is the red- and white-flowering *Salvia microphylla* 'Hot Lips' (little leaf sage), the crimson-blushed new leaves of *Arctostaphylos* 'Sunset' (manzanita), the burgundy and white blooms of *Mimulus* 'Changeling' (monkey flower), and the umbrella-shaped clusters of tiny, red *Achillea millefolium* 'Paprika' (yarrow) flowers.

Red, yellow, and green variegated *Euphorbia* × *Martinii* 'Ascot Rainbow' (spurge), adds touches of yellow that are enhanced by the golden blooms of *Coreopsis auriculata* 'Nana' (dwarf tickseed).

With Coleman's encouragement, Bellefeuille added yet more color, such as orange lilies, brilliant orange *Eschscholzia californica* (California poppy), purple-flowering *Penstemon heterophyllus* 'Margarita BOP' (foothill penstemon) and *Crocus vernus* 'Remembrance' (spring crocus).

To balance the intense colors and rounded shapes, Bellefeuille added masses of soft, mounding silvery blue grasses such as native *Festuca californica* 'Phil's Silver' and *Festuca idahoensis* 'Siskiyou Blue' and 'Stony Creek'.

LEFT: *Mimulus* 'Changeling' is a hybrid of California native monkey flower, perennials with narrow, deep green leaves and lovely monkey-faced flowers that tolerate dry soils. An occasional deep irrigation extends their bloom into summer.

BELOW LEFT: *Penstemon heterophyllus* 'Margarita BOP' is a hybrid of two California natives. Flowers range from sky blue to nearly cobalt, with lavender throats. Though not long lived, this penstemon requires very little irrigation in this well-mulched garden.

BELOW: Bellefeuille used four selections of blue-bladed fescues and other grasses to balance the garden's riot of colorful flowers, including this native monkey flower cultivar, *Mimulus bifidus* 'Esselen'.

Pink Storybook Bungalow

The pink storybook bungalow is a study in designing with a careful eye and restrained hand. Following an extensive remodel, Coleman had the house painted a coral pink similar to the original color. For contrast, she chose bright teal and deep cherry red trim and used tone-on-tone to bring out the architectural details.

For the garden, form comes first, so Bellefeuille mounded dirt to sculpt planting mounds on the otherwise flat 1,400-square-foot space. Then, Bellefeuille set out to match her plant palate to the home's color palette.

Outside the home's picture window, Bellefeuille envisioned a large shrub or small tree lacey enough to let light into the room but dense enough for privacy. A tall, upright dwarf olive tree was the perfect option. Its mass offers privacy while its narrow, silvery foliage lets light in and complements the teal trim.

Next to the olive, Bellefeuille planted an arching, vase-shaped *Cestrum* 'Newellii', a shrub whose cherry red flowers match the front door. The pinks and burgundies are repeated in pink-flowering *Eriogonum grande* var. *rubescens* (native buckwheat), *Epilobium* 'Schieffelin's Choice' (California

The entry walkway of this colorful and beautiful waterwise garden in Santa Rosa, California, is flanked by flowers, while colorful foliage fills the rest of the beds.

The owners accented the home's coral pink stucco with teal blue window trim and a burgundy red front door. Bellefeuille chose plants that echo the home's colors and emphasize California natives, accented with other waterwise Mediterranean-climate plants.

Garden Gallery

From inside the garden, the icy blue-green foliage, yellow blooms, and deep pink roses stand out against the garden's white picket fence.

fuchsia), *Punica granatum* 'Nana' (dwarf pomegranate), *Muhlenbergia capillaris* (pink muhly grass), and red-berried *Heteromeles arbutifolia* (toyon).

To compliment the teal trim, Bellefeuille planted blue-flowering rosemary, small mounds of ornamental blue fescue grasses, and several varieties of glaucus-leaved manzanita.

Bellefeuille knew that a bit of complementary color would brighten the pinks and blues. *Stipa gigantea* (giant feather grass), with its golden flower heads, and the soft yellow flowers of *Eriogonum nudum* 'Ella Nelson's Yellow' (naked buckwheat) add just the right touch.

One of the most exquisite pairings in this garden is a beautiful specimen of manzanita, *Arctostaphylos* 'Lester Rowntree', planted against the coral pink walls. The small blue-green leaves are edged in the same coral pink and the tiny springtime flowers are almost a pearl pink color, similar to the billowy flowers of the Sally Holmes rose that clambers over the white picket fence.

"Designs are form driven and function driven first, and then color is the next layer," says Michelle Bellefeuille. Beyond form and function, this garden is beautiful, colorful, quiet, and demure.

ABOVE LEFT: *Cestrum* 'Newellii' is one of the few large shrubs in this garden. While the garden's palette is mostly soft colors, the cestrum blooms a deep cherry red to match the front door.

ABOVE: *Eriogonum nudum* 'Ella Nelson's Yellow' accidently popped up in a batch of deep pink-flowering native *Erigeron grande* var. *rubescens*. When it bloomed, Bellefeuille and the owners quickly recognized the soft yellow as the perfect complement for other colors in this garden.

LEFT: *Arctostaphylos* 'Lester Rowntree' is a selection of native manzanita with lovely pink flowers that match the home's coral pink stucco. New leaves emerge green with red margins, stems, and veins. As they age, leaves turn a glaucus blue similar to the window trim.

FACTS AND FIGURES

ELEVATION 173 feet
ANNUAL AVERAGE RAINFALL 30 inches, mostly October to April; dry in summer
SUMMER HIGH 87°F, typically in July and August
WINTER LOW 35°F in December and January
HUMIDITY Ranges from 95% in June to 73% in October
IRRIGATION In-line drip
SOIL TYPE Amended loam (from sheet mulching) over heavy clay
USDA ZONE 9b
DOMINANT PLANT TYPE Mixed Mediterranean broad-leaved plants with many California natives

KEY PLANTS FOR COLOR

BLUE BUNGALOW
Achillea millefolium 'Paprika'
Arctotis The Ravers 'Pumpkin Pie'
Ceanothus 'Centennial'
Cistus ×*skanbergii*
Eschscholzia californica
Euphorbia × *Martinii* 'Ascot Rainbow'
Rosa Flower Carpet
Mimulus bifidus 'Esselen'
Mimulus 'Changeling'
Salvia microphylla 'Hot Lips'

PINK STORYBOOK BUNGALOW
Arctostaphylos densiflora 'Sentinel'
Arctostaphylos 'John Dourley'
Arctostaphylos 'Lester Rowntree'
Epilobium 'Schieffelin's Choice'
Eriogonum grande var. *rubescens*
Eriogonum nudum 'Ella Nelson's Yellow'
Muhlenbergia capillaris
Rosmarinus officinalis

Garden Gallery

Native Confetti

While colorful gardens are sometimes a side road along the journey, there are times when they are the destination. In the urban core of Los Angeles there stands a lovely, one-hundred-year-old craftsman home that not long ago was rescued from decades of a terrible peach stucco overcoat. The owners restored its sage green-shingle siding that serves as a backdrop for one of the best native wildflower displays in the city. In the clear, warm California spring sun, the garden explodes in shades of magenta, mango, periwinkle, indigo, scarlet, and lemon. Passersby stop and gape. Cars driving by slam their brakes hard and then back up to take a look.

This home belongs to Mary Beth Fielder and Murray Cohen, and the garden is a collaboration between Fielder and landscape designer Nick Dean. In 2007, Fielder read Douglas Tallamy's *Bringing Nature Home*, a book about how we can better sustain nature and wildlife in our home gardens. She was taken by the author's assertion that nature being "somewhere else" doesn't hold up anymore: "Everything is being developed," she says, "there is no 'somewhere else.' Natural ecosystems are being destroyed," she explains, by development that leads to mass extinctions. The solution is to restore the ecosystem one urban or suburban garden at a time.

The book inspired Fielder to bring nature into her own garden. She set out to replace about 1,000 square feet of front lawn with coastal sage scrub habitat, only

This century-old Craftsman in the heart of Los Angeles has a native garden that is home to native wildlife year-round. In spring, the garden explodes in a colorful confetti of California wildflowers.

What was once a 1,000-square-foot front lawn was transformed with minimal structure, just a decomposed granite walkway and small sitting area, along with some boulders surrounded by a lush tapestry of wildflowers. Now, the garden welcomes neighbors and inspires them to create their own urban habitat gardens.

to realize that this was impractical and perhaps even impossible. So she regrouped and decided on a garden of California natives that would serve two purposes: it would provide habitat for native wildlife while being attractive enough for neighbors to follow suit.

Fielder asked her friend, landscape designer Nick Dean, to create the broad brushstrokes of the garden, which she would then fill in with plants. Dean played off preexisting features that framed the garden: the home itself, the neighbor's tall hedge of deep green Italian cypress along one side of the property, a red-flowering bottlebrush just inside the cypress hedge, and an enormous pine at the driveway entrance.

Dean did a rough drawing of the layout, then switched to drawing with chalk on the ground. He lined out a pathway of decomposed granite and stone to traverse

ABOVE LEFT: Designer Nick Dean planted purple-flowering *Salvia clevelandii* and blue-flowering *Ceanothus* 'Dark Star' beneath a mature red-flowering bottlebrush tree that was already in the garden.

ABOVE RIGHT: Elegant *Clarkia unguiculata* is native to California's coastal ranges. It attracts hummingbirds to the garden and reseeds heavily, adding blooms to the garden year after year.

the gentle slope from sidewalk up to the brick front porch. Part way up the slope, he added a flat landing with just enough room for a table and two chairs. Large rocks punctuate the slope, creating what Dean refers to as a "canvas on which Mary Beth could paint her natives."

In the meantime, Fielder immersed herself in learning about California native plants. For this garden Fielder wanted "hot colors," she says, "yellows, reds, and oranges in one area of the garden, moving into pinks on the other side." Purples and deep blue also made their way into her palette.

Fielder envisioned a color tapestry, best achieved by arranging colors in drifts. "The lilac verbena on the hill," Fielder says, "is a foundation drift of color that is always there." Dean massed purple-flowering *Salvia clevelandii* (Cleveland sage) and blue-flowering *Ceanothus* (California lilac) 'Dark Star' and 'Frosty Blue' to set off the neighbor's deep green cypress.

Low-growing silvery-leaved salvias, such as 'Bee's Bliss', and *Ceanothus gloriosus* var. *exaltatus* 'Emily Brown', along with *Arctostaphylos edmundsii* var. *parvifolia* (bronze mat manzanita), provide an evergreen carpet beneath annuals such as pink and white clarkias and bright orange

California poppies. When the annuals disappear, the carpet remains.

The area surrounding the pine tree was planted as a woodland garden. The backbone is feathery green *Achillea millefolium* (yarrow), which blooms from white to rose pink. Hot pink *Mimulus* 'Trish' (monkey flower) grows amid the yarrow, matching colors through the year.

Flowering annuals light up the parkway planting, too. In fact, if you stand across the street and look back at the house, the sidewalk completely disappears behind the riot of blooms.

Los Angeles receives only 15 inches of rain each year, all of which falls between November and March. This garden's big bloom follows the end of the rains by a month or so. By summer, the annuals are gone, and without irrigation many of the perennials go to sleep. Fortunately, Fielder planned her garden to have color even in the hottest months of the year. "Part of the trick is to pick things that bloom at different times," she says. "Later in summer, buckwheat makes flowers. Monkey flower takes a rest in summer but bloom in waves the rest of the year." Lemony yellow *Encelia californica* (California bush sunflower) blooms through summer, while hot red *Penstemon eatonii* (firecracker penstemon) and the lavender- to rose-blooming

ABOVE LEFT: *Mimulus* 'Trish' is the perfect contrast to a background of silvery, feathery, and aromatic *Artemisia californica* (California sagebrush).

ABOVE RIGHT: *Dudleya pulverulenta* (chalk liveforever) provides the backdrop for one of the many varieties of monkey flower planted in this garden.

Penstemon heterophyllus 'Margarita BOP' (foothill penstemon) add to the garden's year-round rainbow.

Fielder waters once a week through the heat of summer, but not much more than that. Her garden's clay soil holds moisture for a long time, so it can be a challenge to water enough, but not so much that plant roots drown.

In terms of maintenance, caring for this garden feels more like therapy than chore. "I get such pleasure from this garden," Fielder says. "When I get home from work I spend 15 minutes walking through the garden and picking a weed here and there. I don't mind cutting it back when it's time. I get such satisfaction from seeing how I've created habitat for butterflies and birds."

FACTS AND FIGURES

ELEVATION 178 feet
ANNUAL AVERAGE RAINFALL 15 inches
ANNUAL RAINY DAYS 22.5 days between November and March; dry in summer
SUMMER HIGH 85°F, typically in August
WINTER LOW 47°F in December and January
HUMIDITY Ranges from 90% (July to August) and 58% (December)
IRRIGATION Started with overhead, changed to drip
SOIL TYPE Clay
USDA ZONE 10a–10b
DOMINANT PLANT TYPE California native

KEY PLANTS FOR COLOR

Achillea millefolium
Ceanothus 'Dark Star' and 'Frosty Blue'
Ceanothus gloriosus var. *exaltatus* 'Emily Brown'
Eriogonum grande var. *rubescens*
Mimulus 'Jelly Bean' hybrids
Penstemon eatonii
Penstemon heterophyllus 'Margarita BOP'
Salvia 'Bee's Bliss' and 'Pozo Blue'
Salvia clevelandii 'Compacta'
Salvia spathacea
Verbena lilacina 'De la Mina'

Garden Gallery

Color at the Coast

Dustin Gimbel, one of Southern California's up-and-coming landscape designers, has a very impressive pedigree. As a teenager, he worked for the late Mary Lou Heard, a noted Southern California plantswoman. After earning a horticulture degree from California State Polytechnic University, Pomona, in 2002, Gimbel embarked on a series of horticultural internships across the globe.

Gimbel has worked with the best, including Dan Hinkley at Heronswood and Christopher Lloyd at Great Dixter in England. After earning the Royal Horticultural Society's Wisley Diploma in Practical Horticulture, Gimbel returned home to California for a brief vacation. When he rediscovered what it felt like to live in year-round sunshine, he decided to stick around.

Gimbel designs gardens along California's south coast. His own garden surrounds a pistachio-colored 1930s bungalow only two miles in from the surf in the city of Long Beach. The home's front garden is small, just 60 feet deep by 150 feet wide, and enveloped in a "green wall" of evergreen fig (Ficus nitida). The tall hedge buffers noise and offers privacy from the neighborhood traffic. And, Gimbel says, he likes living "in a big green box."

Blue, orange, burgundy, silver, and black form the color palette in Gimbel's front garden, starting with a tall *Acacia pendula* (weeping acacia) that towers over a walkway that cuts a diagonal across the garden. The tree's coppery brown bark is

In this small garden, Gimbel laid a diagonal "boardwalk" made from wood scraps. Silver-leaved *Acacia pendula* at the far end pulls the eye past golds, purples, blacks, chartreuse, and orange tones along the way.

73

Leucospermum 'Veldfire' (South African pincushion) blooms intense yellow and orange, and has bright green foliage that stands out against the pistachio mint green of Gimbel's beach bungalow. Black accents include succulent *Aeonium* 'Zwartkop' and spiny *Dyckia* 'Black Gold', whose orange flower spikes echo the pincushion flowers. Other orange tones come from *Euphorbia milii* 'Apricot' (crown of thorns) and succulent groundcover *Sedum nussbaumerianum*.

Garden Gallery

Gimbel's Long Beach, California, "lawn" consists of *Frankenia thymifolia* accented by succulents plus *Dyckia* 'Black Gold', a very spiny bromeliad. He hand waters this small coastal garden at most once every two weeks in summer.

topped in silver, blue-gray leaves. Beneath the tree is a tall ceramic pot glazed in the same copper brown as the acacia bark. Wispy fronds of *Russelia* 'Night Lights Tangerine' (firecracker plant) spill out and over the pot, each tipped in a tube-shaped, sherbet orange flower.

Gimbel is all about creating mood in the garden. "Its about excitement and contrast and little surprises," he says. Black in particular seems to catch his eye, including the dramatic burgundy-almost-black foliage of *Euphorbia* 'Blackbird'. Next to a miniature lawn of the tiny-leaved, creeping groundcover *Frankenia thymifolia* (sea heath), Gimbel planted *Dyckia* 'Black Gold'. Dyckias are armored bromeliads with purple, silver, or black blades that are silvery

Garden Gallery

Gimbel painted one outer garden wall deep gray, then added chartreuse, yellow, silver, deep green, and burgundy plants: silver-leaved and yellow-blooming *Achillea* 'Moonshine', feathery chartreuse *Coleonema* 'Sunset Gold', and *Melianthus major* 'Purple Haze', whose broad fronds are blush purple at the midrib. The studded cement orbs were cast in concrete from rubber pet toys Gimbel found in a store. He added a rubber yellow one to the collection, just for fun.

OPPOSITE: Gimbel makes stacked screens of hypertufa stones. Their rounded forms and soft gray colors counterbalance the bright intensity of the plants.

underneath and have stalks of golden yellow or orange flowers. "From one angle you don't see the dyckia at all," Gimbel says, "but then you turn around and there it is." Similarly, he underplanted an almost black-leaved succulent, *Aeonium* 'Zwartkop', with bright orange *Sedum nussbaumerianum* (coppertone stonecrop).

This intensity is balanced with a nearby cluster of plants all in soft greens and silvers that give the space a restful aura. It features many succulents, such as *Agave attenuata*, *Echeveria gigantea* (giant hen and chicks), and *Cotyledon orbiculata* var. *oblonga* 'Flavida' (finger aloe), along with broad-leaved plants such as fringy silver-green *Adenanthos sericea* (coastal wooly bush). The plants play off stacked "screens" of rustic, round gray "stones" that Gimbel makes of hypertufa (concrete mixed with perlite). The net effect? "When your eyes hit the combination of colors and textures it creates excitement," Gimbel explains, "then your eye jumps away and then jumps back."

Broad-leaved plants mix easily with succulents that require no more than hand watering once every week or two, even in the heat of summer ("and I probably water too much," Gimbel says). As much as the succulents contribute to the garden, "it would be boring to do 100% succulents," he says. "You just can't get all the textures you need…. I want it to be low water but not look like a desert garden."

Gimbel's rear garden has an entirely different color palette and offers an entirely different experience. Part of the space is a small nursery where he propagates hard-to-find plants. He has a toolshed,

Gimbel's so-called deconstructed screen is meant to act like a "fence that someone came back and added windows to." He also added elements to traverse the screen, like this cinderblock pond with its upright papyrus and yellow glass floats.

Gimbel haunts second-hand stores for odd objects, which he then casts in soft gray concrete and places carefully among the plants.

Gimbel plays with shapes and textures, as well as colors. Here, the focus is on burgundy-almost-black played off bright greens and silver grays, along with touches of burnt orange. In this coastal garden, all of these plants thrive with almost no irrigation.

Garden Gallery

ABOVE: Reflections in the garden: shiny yellow glass floats and still water at sunset.

ABOVE RIGHT: Gimbel layers his colors; the pistachio mint green of his bungalow's wood siding is echoed in the leaves of a fancy, succulent echeveria planted at the base of the purple-brown patterned bromeliad *Hohenbergia correia-araujoi*. This is classic Gimbel artistry in color and structure.

beehive, vegetable garden, and an outdoor dining room—all in a 20 by 30 foot area.

Here, the color palette is yellow, gold, purple, burgundy, and gray. Gimbel applied hardy board to a wall he shares with a neighbor, then painted it gray. He painted the backside of the detached garage dusky purple to "be sympathetic" with the gray. Where the two walls intersect is a bed of golden yellow *Coleonema* 'Sunset Gold' (breath of heaven) combined with purple-blush *Melianthus major* 'Purple Haze' (dwarf honeybush), yellow-blooming *Achillea* 'Moonshine' (yarrow), wine-bladed *Cordyline* 'Festival Grass', yellow-orbed *Craspedia globosa* (Billy buttons), and burgundy-leaved *Dodonaea viscosa* 'Purpurea' (hopseed bush).

In spring, the purple wall is nearly hidden by the vining *Aristolochia gigantea* (giant Dutchman's pipe) that in another climate zone would need much more irrigation than here, so near the coast. By summer, the vine is covered with its enormous and slightly erotic looking burgundy purple flowers.

Closer to the house, Gimbel erected what he calls a "deconstructed screen" to define a room within the room. "The idea is that it's like a fence that someone came back and added windows to," he says, "so there are nice framed views into other garden spaces. It creates an intimate dining area without losing the views and sense of space." The connection between inside the dining area and outside is made in part, by a water feature—not quite a pond—built of gray cinder blocks, that transects the screen.

The gray cinder block echoes the gray garden wall, while yellow glass orbs that bob in the water pick up the colors of the plants behind. Bright golden California poppies bloom around the cinderblock in spring.

Gimbel explains his color choices, "Golden colors are exciting. There are many cloudy day at the beach. The gold looks great in morning and in afternoon when it's cloudy. Burgundies balance that." His color combinations are fantastic, if not unusual. When asked about them, he pauses for a moment. "Most of the stuff is instinctual. I don't think about it, I just do it," he says. "I spent a lot of time in gardens in England. I would find a bench and just sit…. It's about creating experience. The more defined or controlled the colors are, the more you can predict what someone is going to get out of the experience."

"People should be more willing to try new things in the garden," Gimbel continues, "and not to fear failure. If a plant dies it's an opportunity to try something else…. It's the same thing for color combinations. Try something you may not expect to like. You can always change it."

FACTS AND FIGURES

ELEVATION 25 feet
ANNUAL AVERAGE RAINFALL 14 inches between November and March; dry in summer
ANNUAL RAINY DAYS 21
SUMMER HIGH 81°F, typically in August
WINTER LOW 49°F in December and January
HUMIDITY Between 72% (November) and 88% (September)
IRRIGATION Hand watered at most once weekly or every two weeks in the heat of summer
SOIL TYPE Silty loam
USDA ZONE 10b
DOMINANT PLANT TYPE Mixed Mediterranean

KEY PLANTS FOR COLOR

Achillea 'Moonshine'
Aeonium 'Cyclops' and 'Zwartkop'
Agave attenuata
Dyckia 'Black Gold'
Echeveria gigantea
Euphorbia 'Blackbird'
Euphorbia cotinifolia
Euphorbia milii 'Apricot'
Leucospermum 'Veldfire'
Melianthus major 'Purple Haze'

Garden Gallery

Jewel-Toned Mediterranean

Cerise, ochre, gold, terra cotta, burgundy, and emerald—the colors of San Diego garden designer Amelia Lima's garden are as rich as her background. Lima was born and raised in Brazil, in an environment flavored by the famous landscape architect, artist, and botanist Roberto Burle Marx. Marx designed some of the world's most notable public spaces, including the Copacabana Promenade in Rio de Janeiro, a 2½-mile-long beach wave tile mosaic sidewalk.

As a young woman in the late 1980s, Lima spent a week in Burle Marx's studio, soaking up the environment, roaming the extensive grounds, and observing the artist, who was in his eighties at the time. Burle Marx was said to have "painted with plants," and experimented extensively with native and exotic plants to see what best suited Brazil's climate.

From Burle Marx, Lima learned to understand the climate and the limitations of each site, and then match plants appropriately. "Roberto looked for plants indigenous to the area, that were happy to be there and weren't fussy or needed to be

The colors of San Diego garden designer Amelia Lima's garden are as rich as her Brazilian background.

pampered," she says. He also traded seeds with people around the world to find exotics that would grow easily in Brazil. His grounds were filled with plants. "I was so impressed by his garden," Lima recalls. "It was his lab, where he would try his ideas, his plant combinations, and also where he grew plants for his projects."

In Burle Marx's time, there were no nurseries that grew the plants he wanted to use; seemingly no one in Brazil valued the bromeliads and other regional natives. Instead, he collected native plants on-site before demolition began, propagated them, and grew them in his shade houses to be used in his designs. "The plants he used, no one at the time thought they were worthy of a garden," Lima says, "today, that is amazing."

In her professional practice as a landscape designer, Lima uses the lessons she learned from Burle Marx, with her own garden in San Diego being the best example. When Lima redesigned the tiered landscape that came with the property, she started first with the climate and with local limitations. "Water is a precious resource here," Lima says. "A garden that consumes a huge amount of water would never be at peace in this environment."

The garden would be low water, then, but what about the style? Lima invoked Burle Marx's love of big, bold, architectural statements. "His gardens had to be pictorial," she says, "so that even when viewed from far away they make an amazing picture. That is why they photograph so well, they are actually paintings"—paintings made with plants.

To achieve that effect, Lima designed a low, terra cotta–colored stucco wall that divides the sloped expanse horizontally,

Lima's garden renovation began with an assessment of the climate and local limitations. A garden of thirsty plants was out of the question because, as Lima says, they "would never be at peace in this environment." Instead, she used plants from Australia, South Africa, and other regions that share San Diego's Mediterranean climate.

creating an upper private garden and a lower streetside garden. Broad stairs that center on the home's large, arched front window connect the sidewalk to the upper entry walk.

When it came to plants, Lima concentrated on shape and form. "To me, the beauty of these plants is their architecture alone," she says. "If you took a black and white photo of my garden, it would still be beautiful because of the architecture. When you add color, it becomes a sundae with all the sprinkles on top. That is the way I approach gardens."

"I love color, I love color," Lima continues, and like Burle Marx, she prefers colorful foliage to colorful flowers. "I look at people who know how to use color, like Luis Barragán in Mexico…. The way he uses bright colors on walls, the way the plants create shadows against colored walls, it creates emotion and color is about emotion."

That interplay of colored walls and colorful foliage characterizes this garden. The exterior of Lima's home is soft gold with terra cotta roof tiles. Lima anchored the garden with a tall green-leaved, lemon-scented *Leptospermum petersonii* (tea tree) on the driveway side of the front stairs. On the far opposite corner of the house she kept a 12-foot-tall, burgundy-leaved *Cordyline australis* 'Red Star' (red grass tree), one of only three or four plants that remain from the garden's previous incarnation. Taken together, the two plants frame the garden.

In between and surrounding those framing elements, Lima used big, bold plants. A pair of large Mexican grass trees (*Dasylirion longissimum*) dominates one corner of her property. In the afternoon

Lima shows her deep love of color through designs based on colorful foliage. Colorful flowers are "here today, gone tomorrow," she says.

breeze, the long, narrow, horizontal green blades wave past each other, creating a captivating optical illusion.

Though Lima's plant colors are an accident of their architecture, what a fortunate accident they are! She combines colorful succulents with the bright-colored foliage from Australian and South African shrubs. Rounded waves of *Senecio mandraliscae* (blue chalk fingers) hug the sidewalk, and fill in the spaces between upright, multicolor *Leucadendron* 'Wilson's Wonder' (conebush), with its green, gold, and red scalelike blades. There are fabulous agaves such as the variegated green and cream *Agave americana* var. *mediopicta* 'Alba' (white-striped century plant), traditional green *A. attenuata*, and variegated *A. attenuata* 'Kara's Stripes'. The unexpectedly teal blue *A. attenuata* 'Nova' is reminiscent of giant blue cabbage.

Masses of green *Aeonium urbicum* rosettes combine with the coral and yellow blooms of succulent bladed *Aloe saponaria* and *A. striata*. Colors echo and weave the scene together. Lima is particularly fond of *Sedum nussbaumerianum* (coppertone stonecrop), whose knobby orange succulent blades light up just about any other color in the plant palette.

None of these plants were chosen for their flowers. "People give too much value to flowers," Lima explains. "To me flowers have no value. If you blur your eyes, flowers are just little specks on top of a plant—here today, gone tomorrow, and just so boring. Blur your eyes and look at an agave, it's still there. When plants have a strong architectural shape *and* a rich color, you have everything in the same place."

"There are some colors I am passionate about," Lima continues, " limey green,

In the afternoon breeze, the long, narrow blades of a pair of large Mexican grass trees wave past each other, creating a fascinating optical effect.

A study in upright colored foliage: (left) *Beschorneria yuccoides* (Mexican lily), which is an agave relative with bluish blades and tall coral flower stalks, (bottom center) variegated green and gold *Agave attenuata* 'Kara's Stripes', and (right) the olive, wine, and bronze leaves of *Phormium* 'Dusky Chief' (New Zealand flax).

Lima's love of architecture in the garden works on both macro and micro scales, as seen here in the coppery brown teeth of *Agave parryi* var. *truncata*, whose green blades show the embossed images from other leaves pressed tightly together as they developed in the center of the agave rosette.

yellows, and oranges. Blue I find combines in the garden very well. It is an easy color that goes with everything…. greens, magentas…. I *love* chartreuse in the garden. We don't have a lot of plants that are chartreuse."

Many of Lima's favorite colors show up in her container plantings. Here, Lima's designs are bountiful. "I start with the container," she explains. "The container size dictates the size of the arrangement." She favors plants that are twice the height of

This succulent symphony includes the diminutive blue-bladed *Senecio serpens* (blue chalksticks), orange-bladed *Sedum nussbaumerianum*, and *Aloe striata* (coral aloe) in bloom, all framing a large *Agave* 'Blue Flame'.

the container. In the final arrangement, then, the container makes up the lower third while the plants comprise the rest. Lima's containers are planted to overflowing. "I bring the soil all the way up—leaving only a half-inch from the rim," she adds, "and use plants that cascade down."

Lima likes containers that are limey green or blue; "I love terra cotta pots. I love pots with architecture and nice shapes. Pots that have textural architecture like beautiful bowls that have lines in them." Even empty pots have a place in Lima's designs: "Empty pots in the garden are focal points, but if you put an empty pot into the garden, the pot has to have character, after all, they are architecture."

Lima's soil is sandy so it drains well. She uses in-line drip irrigation that runs for 22 minutes every time it runs. In spring and fall, she irrigates every zone once every eight days. In summer, a reflective coating on the large front window exacerbates the intense heat of the south-facing front garden, so at that time of year she runs her irrigation once every six days. In winter, the system runs only once in ten days unless there is a long dry period. Winter irrigation

Lima puts an emphasis on color and emotion, which separate her vibrant garden from the typical landscape.

is augmented by runoff from the roof. The front downspout connects to a buried perforated pipe that runs the length of the garden, parallel to the house. Rainwater that falls onto her roof flows into her garden's sandy soil.

Lima mulched the entire garden when it was first planted. For a short while she replaced the mulch on a regular basis, but after just three years, the plants had grown to the point that there wasn't room for mulch.

Low maintenance, low water, high color, and high emotion are the elements that make this garden so stimulating. As Lima says, "Color is about emotion. If there is no emotion, it is not a garden, it is just hardscape."

When planting containers, Lima matches the size and shape to the plants. This container holds a stately jade.

FACTS AND FIGURES

ELEVATION 224 feet
ANNUAL AVERAGE RAINFALL 13 inches between October and March; dry in summer
ANNUAL RAINY DAYS 24
SUMMER HIGH 84.5°F, typically in August
WINTER LOW 43°F in December to January
HUMIDITY Ranges from 68% in October and December to close to 90% in January, April, September, and November
IRRIGATION In-line drip
SOIL TYPE Sand
USDA ZONE 10a
DOMINANT PLANT TYPE Succulents mixed with dry-growing Mediterranean plants

KEY PLANTS FOR COLOR

Aeonium 'Kiwi'
Anigozanthos 'Bush Dawn'
Crassula capitella 'Campfire'
Euphorbia tirucalli 'Flame'
Kalanchoe luciae
Leucadendron salignum 'Winter Red'
Sedum nussbaumerianum
Senecio mandraliscae

Garden Gallery

Northern Lights

Two squares of tired lawn, two liquidambar trees, and one saucer magnolia; a brick walkway and a small, undistinguished 1950s ranch-style home in a subdivision of similar homes, circa 1950s—these are the elements that made up Michelle Hornberger and Terry Short's garden when they purchased this Sunnyvale, California, home in 1993. "I never liked how the front yard looked," says Michelle Hornberger. "I tried to dress it up with pots and yard art. The mow and blow guys kept it looking neat but I never wanted to spend any time out there." With the price of property in the neighborhood skyrocketing, Hornberger thought it a waste for the front yard to be such an unattractive and underused space.

Today, the front yard sports a garden of contrasting deep and bright tones and textures that cover about 1,000 square feet. It features a chocolate brown, zigzag wall that defines an outdoor seating area adjacent to the front porch. The plants in this garden require little more than the 20 inches of rain that fall from the sky in a typical year.

Landscape architect Leslie K. Dean, principal of Dean Design Landscape Architecture, designed this garden in response to Hornberger's request for plantings

Updated color, a reconfigured front entry, and a redesigned garden transformed a previously undistinguished 1950s ranch-style home in Sunnyvale, California.

The homeowners asked landscape architect Leslie K. Dean for a garden full of colorful, textural plantings that could be maintained without excessive watering. This kangaroo paw (*Anigozanthos*) fits the bill beautifully.

ABOVE: *Asteriscus maritimus* is a low-growing waterwise perennial whose mounds edge the garden's beds.

BELOW: Plants like *Lavandula stoechas* 'Wings of Night' (Spanish lavender), *Hemerocallis* 'Lady Eva' (daylily), *Phormium* 'Bronze Baby' (New Zealand flax), red-flowering salvia, and yellow *Asteriscus maritimus* (gold coin aster) meet the owners' request for color and diversity without the need for much irrigation.

Garden Gallery

The homeowners' original wish list included an old-fashioned front porch. When that proved impossible, Dean designed a seating patio just outside the porch, complete with planting pockets tucked into the zigzag pattern of the wall.

that use less water while still offering lots of color and texture—Hornberger also wanted a lime tree.

Because of their years-long friendship, Dean knew Hornberger as a hands-on gardener who likes variety. At the same time, too much variety can become chaotic, so there needed to be an element of cohesion to the new garden.

The original house included a small covered porch—more stoop than porch—defined by a chest-high brick planter wall. A narrow, brick path led straight from door to sidewalk. The planter and brick came out, courtesy of Terry Short and some elbow grease, to be replaced by sleek, poured concrete steps that lead to a new concrete walkway. Deep-tone slate covers the concrete by the door. Bands of the same slate are set into the walkway.

Hornberger's wish list included an old-fashioned front porch, but the home's front porch was too small for that. Instead, Dean designed a seating area a few feet out from the entry. Walls set 3 feet high in a zigzag pattern create a cozy nook for a pair of cushioned chairs and a table. Outside the wall, the zigzag forms strategically placed planting pockets.

As Dean designed the garden, Hornberger worked on a new paint palette. She favors deep, warm colors, so with some input from Dean, her final choice was mocha siding with creamy white for the trim. A chocolaty brown color went onto the stucco wall surrounding the seating area. The color palette pulls together the deep tones of the slate front porch, the home's aged brick, and the gold-colored decomposed granite Dean used in the seating area.

The plants pop against the deep paint tones. The foliage of *Phormium* 'Bronze Baby' creates a tone-on-tone effect while adding hints of red and olive. Its bold, upright structure also anchors the new landscape. Dean placed a lime tree in the planting pocket on the backside of the seating wall. Its bright green leaves shine next to the chocolate brown.

Year-round perennials provide most of the color in this garden. Burgundy-flowering *Pelargonium sidoides* (South African geranium) offers contrast to the bright pink flowers of *Penstemon* 'Apple Blossom' and *Salvia* 'Plum Wine'. Vining *Lonicera sempervirens* 'Major Wheeler' (trumpet honeysuckle) adds its cerise, apricot, and gold tubular flowers that draw hummingbirds.

Lavenders, purple-flowering *Salvia leucophylla* (Mexican bush sage), catmint (*Nepeta*), and the almost-blue flowers of *Geranium* 'Rozanne' add another dimension of color to the garden. The lavenders' gray foliage and the silver foliage from *Rhodanthemum hosmariense* (Moroccan daisy), contrast with the phormium's bronze foliage and red-leaved *Berberis thunbergii*

Garden Gallery

Dean chose some plants for their color contrasts; others, like *Phormium* 'Bronze Baby', were chosen for their structure and for a tone-on-tone effect, as seen here in the hints of red and olive.

The bright pink, apricot, and yellow flowers of *Lonicera sempervirens* 'Major Wheeler' are a beautiful contrast to chocolate brown garden walls.

Three kinds of lavenders bloom in this garden and perfume the air with their resinous fragrances.

Long-blooming *Geranium* 'Rozanne' is a bit to the thirsty side, but this garden is irrigated with in-line drip irrigation installed 4–6 inches below the surface. In practice, the garden needs only periodic irrigation from May through September.

var. *atropurpurea* 'Crimson Pygmy' (barberry). Against the chocolate and mocha walls, all of these plants look stunning.

The only plants remaining from the original garden are a tall liquidambar tree in the parking strip and *Magnolia* ×*soulangiana* in front of the bedroom wing. The magnolia loses its leaves in the winter, then erupts in enormous, rose blush–colored flowers in early spring. Dean blanketed the ground beneath the magnolia with pink-flowering wild strawberry (*Fragaria* 'Pink Panda'). The pale white, rose, and pink colors glow against the deep tones behind it.

The new garden has so much to offer that summer mornings find Hornberger and Short there, drinking coffee, while summer afternoons are spent enjoying the afternoon shade. Neighbors come by to chat and comment on their colorful, beautiful new garden. "We all love it," Hornberger says.

FACTS AND FIGURES

ELEVATION 79 feet
ANNUAL AVERAGE RAINFALL 20 inches, between December and February; dry in summer
ANNUAL RAINY DAYS 40
SUMMER HIGH 79°F, typically from July to September
WINTER LOW 42°F in December and January
HUMIDITY Ranges from 60% in January to 90% in September and December
IRRIGATION Strawberry groundcover irrigated by overhead spray; all other areas are in-line drip, installed 4–6 inches below the surface; irrigation runs from May through September only
SOIL TYPE Adobe clay
USDA ZONE 9b
DOMINANT PLANT TYPE Mixed Mediterranean trees, perennials, shrubs

KEY PLANTS FOR COLOR

Anigozanthos species
Eriogonum grande var. *rubescens*
Lavandula ×*intermedia* 'Grosso'
Lavandula stoechas 'Wings of Night'
Pelargonium sidoides
Salvia leucantha

Garden Gallery

The Color of History

San Luis Obispo is a lovely college town along California's central coast. Its human history goes back to the native Chumash, who hunted, gathered, and fished the area's rich natural resources. Spanish explorers arrived in 1760, followed soon after by Franciscans who built Mission San Luis Obispo de Tolosa in honor of Saint Louis of Anjou, the bishop of Toulouse, France. Since the Spaniards arrived, San Luis Obispo has been ranched and farmed by early Californios, settled by easterners, then colonized by students.

In an older part of town, a bungalow-style farmhouse was recently restored and its garden reborn thanks to new owners, Michael and Theresa Mulvhill, who were attracted to the area because of its rich history.

They call it the "Anholm garden," named for the Dutchman who settled here and farmed this former mission land

This restored farmhouse stands on former mission land in an older neighborhood of California's San Luis Obispo. The home was expanded, and its garden was revived, by new owners.

Garden Gallery

"Signature Hill" is the nickname given to the mounded bed at the center of the garden. Frank and his crew set boulders on the mound to echo those on the nearby mountain.

starting around 1915. The property backs onto a creek at the base of San Luis Mountain. San Luis is one of the "nine sisters," a series of abrupt volcanic peaks that dot the region.

From the street, the modest-looking home has an equally modest (in scale) front garden, about 30 feet deep by 50 feet wide. The rear garden is twice as deep, and because it appears to merge with the mountain behind, it looks much larger.

The property's heritage was part of what attracted Michael and Theresa Mulvhill in 2010. "I wanted to somehow unearth the history of this piece of property," recalls Michael Mulvhill, "it was the landscape itself that called out to me."

While the greater landscape called Michael Mulvhill's name, the gardens were a mess. Landscape designer and contractor Gabriel Frank first visited the property before the purchase was complete. "The backyard looked like someone's grandmother's garden that had been let go for 20 years," he says. There were sweet peas that had reseeded, old roses gone wild, and way too much brick.

A backyard inventory also uncovered an ancient wisteria and remnants of a vineyard the homeowners suspect dates back to the mission era. In front, established old camphor and *Melaleuca* trees were the only plants worth saving.

Coming from Michigan, the Mulvhills found gardening in California's Central Coast area to be an entirely new experience. Frank recalls how they quickly took to the eclectic nature of California gardens. "They wanted a mixture of classic elements with palms, succulents, fig trees, avocado, lemons, and a vegetable garden," he says. At the same time, they wanted their new landscape to be sustainable and drought tolerant.

For this tall task, Frank pulled in his colleague, artist and plantsman Nick Wilkinson.

The men surveyed the property to assess opportunities such as the borrowed view of San Luis Mountain and challenges such as the site's heavy clay soil.

They decided to create large planting mounds both to enhance the mountain view and to counter the clay soil. Frank's crew brought in well-draining soil they used to create a series of 1- to 2-foot-tall mounds at various points in the garden. Plants have an easier time rooting into soil that drains well, at least early in their establishment. Eventually, the roots reach the lower layer of clay, at which point, Frank says, they tap into deeper water and nutrients. The mounds' elevations and contours also create opportunities to set plants at different heights and orientations for layers and drama.

The crew set boulders into the backyard's large central mound to echo the rock outcroppings on San Luis Mountain in the distance. Frank says, "We call the mound 'Signature Hill.' It's how we brought the feel of the mountain into the garden."

The backyard garden is surrounded on three sides by structures. A classic one-car garage, as well as the backside of the home with its wrap-around porch, are painted mocha and green-tint khaki. A refurbished backyard shed is barn red, and all the structures feature white trim. Behind the shed, the mountain is bright green in all but the hottest months of the year when its native cover turns gold and khaki, too.

The colors of the structures are reflected in the colors of the garden. Frank and Wilkinson chose warm gold-tone flagstone and rounded gravel for pathways, mulch, and stone accents. The main plant color palette is yellow, red, and green. Plant color, Frank says, starts with foliage: "I use a lot of striking color in foliage and in variegation. The most valuable component in a garden is a backbone of bold foliage to keep things interesting year-round, always performing, creating contrast."

Toward that end, Signal Hill features a pair of tall *Cordyline australis* 'Torbay Dazzler' (grass tree), which have creamy yellow-, green-, and burgundy-striped leaves. A trio of large green and yellow *Agave vilmoriniana* 'Stained Glass' (variegated octopus agave) balance out the tall grass trees, along with yellow-flowering *Beaucarnea pliabilis* (ponytail palm), deep red-leaved *Cordyline* 'Design-A-Line Burgundy', and golden-bladed *Anemanthele lessoniana* (pheasant's tail grass). Signal

The backyard's color palette compliments the surrounding patio and garage walls, which are mocha and green-tint khaki, along with a barn-red backyard shed.

A shady "oasis" zone of slightly thirstier plants sits under the eaves of the new covered back porch. Green is the dominant color here, thanks to *Cyperus papyrus* (tall papyrus), *Phoenix robelini* (pygmy date palm), and the developing leafy rosette tower of *Echium wildpretii* 'Tower of Jewels'. These are underplanted with drought-tolerant plants that visually tie this zone to the plantings out in the full sun.

Hill is edged in patches of yellow-blooming *Calylophus drummondianus* (sundrop).

More red-, green-, and yellow-colored succulents fill the surrounding beds, including *Aloe cameronii*, whose green blades turn brick red in sun and heat, yellow-flowering *Kniphofia* 'Malibu Yellow' (red hot poker), and *Agave lophantha* 'Quadricolor', a small agave whose yellow and green variegated blades are edged with deep red teeth.

The succulents are offset by broad-leaved plants like red, green, and yellow variegated *Leucadendron* 'Jester' (conebush), burgundy-leaved tree *Agonis flexuosa* 'After Dark', and red- and yellow-flowering perennial *Gaillardia* 'Fanfare Blaze' (blanket flower).

To accent the main color palette, Frank and Wilkinson added touches of deep purple from plants like the very spiny *Dyckia* 'Jim's Red', which blooms brilliant orange, and silver-purple *Dyckia* 'Precious Metal', which blooms red. Blues and silver blues come from *Brahea armata* (blue hesper palm), *Agave* 'Blue Flame', and the desert

Signature Hill features a pair of tall, variegated *Cordyline australis* 'Torbay Dazzler' underplanted with trunkless, red-leaved *Cordyline* 'Design-A-Line Burgundy' and softened by tall tufts of gold and green *Anemanthele lessoniana*. The bed and surrounding beds are edged in patches of yellow-blooming *Calylophus drummondianus*.

Succulents and broad-leaved plants were chosen first for their foliage shapes, structures, and colors, such as red, green, and yellow variegated *Leucadendron* 'Jester', which Frank set next to yellow-flowering *Kniphofia* 'Malibu Yellow'.

The very spiny terrestrial bromeliad *Dyckia* 'Jim's Red' adds some deep purple and blooms brilliant orange.

native *Yucca rostrata* (beaked yucca). In the winter, succulent *Aloe speciosa* (tilt-head aloe) puts out tall spires of coral colored buds that open to creamy white flowers. The spiral stacked buds open in sequence from bottom to top, creating a two-tone effect that plays off the aloe's succulent blue green blades.

Meanwhile, bright purple–flowering *Bougainvillea brasiliensis* arches over the doorway of the red shed, while the sculptural, blue-green bladed *Yucca linearifolia* flanks the shed's front door.

There is no lawn in this backyard. When the couple moved from Michigan, Michael Mulvhill decided he was done with mowing, and lawns were not sustainable in California, anyhow. Still, the new garden includes a "Hollywood strip," a strip of "grass" that runs down the middle of a singlewide driveway. Frank suggested the grass to break up the long expanse of concrete. Michael Mulvhill thinks of it as a nod to the grass he left behind in Michigan as well as an ode to his love of historic California. Rather than plant traditional turfgrass

The magenta-purple flowers of *Bougainvillea brasiliensis* start to emerge in early spring and arch over the shed door by summer. Next to the bougainvillea, yellow-blooming *Beaucarnea pliabilis* serves as the punctuation mark. Beneath it is *Agave vilmoriniana* 'Stained Glass' and the voluptuous rosettes of *Agave* 'Blue Flame'.

On the shady side of the house, an old-fashioned "Hollywood strip" driveway is planted with waterwise *Carex glauca* edged in a mixture of *Aeonium* rosettes combined with the tentacle-like arms of *Asparagus densiflorus* 'Myers' (foxtail fern).

Spaces between the garden mounds act a bit like bioswales, so drier-growing plants are placed closer to the tops of the mounds, while those that appreciate more water are placed near the bottom. Rain chains direct roof water to the soil, though most runoff is collected in a large cistern at the rear of the property.

down the center, however, Frank opted for unthirsty *Carex glauca* (blue sedge).

San Luis Obispo, like the rest of Mediterranean California, gets little rainfall, only 19 inches on average and much less in a drought. But when it does rain in fall, winter, and early spring the Mulvhill's neighborhood tends to flood, so controlling and directing water is a big concern.

In this garden, the spaces between mounds act like bioswales. Frank and Wilkinson sited the more dry-growing plants toward the tops of the mounds; those that appreciate slightly more water are planted toward the bottom.

Rainwater from the roofs of the garage, shed, and main house is directed to a 1,600-gallon cistern beneath the deck at the very rear of the garden. That water is pumped into the garden's in-line drip irrigation system. While this is a very water-thrifty garden, the water in the cistern doesn't last more than a few weeks according to Michael Mulvhill, who adds that if they had more property they would have installed a larger system.

As the garden has developed, the Mulvhills' appreciation has also grown. "I'm continuously amazed by the changes and how subtle they are," Michael Mulvhill says. "Every four or five days, the garden is different." Part of the change is in the color: "if you take your time and observe," says Mulvhill, who looks at the garden through an artist's eye, "you see a lot of color and a lot of changes through the year, like how the blue of the agaves fades, then grows rich, all depending on the play of light." He especially likes to photograph the garden toward the end of day, when the tips of the agave blades turn red and the veins look mauve.

FACTS AND FIGURES

ELEVATION 204 feet
ANNUAL AVERAGE RAINFALL 19.26 inches falling between November and March; dry in summer
ANNUAL RAINY DAYS 32
SUMMER HIGH 83.4°F, typically in August
WINTER LOW 38°F in December and January
HUMIDITY Ranges from 98% in June to 62% in December
IRRIGATION Drip
SOIL TYPE Clay
USDA ZONE 9b
DOMINANT PLANT TYPE Mixed Mediterranean

KEY PLANTS FOR COLOR

Agave 'Blue Flame'
Agave vilmoriniana 'Stained Glass'
Aloe cameronii
Aloe speciosa
Bougainvillea brasiliensis
Calylophus drummondianus
Cordyline australis 'Torbay Dazzler'
Cordyline 'Design-A-Line Burgundy'
Dyckia 'Jim's Red' and 'Precious Metal'
Leucadendron 'Jester'
Yucca 'Bright Star'

"The garden is like art," he continues, "the more you look at it, the more you understand it. A landscape is a lot like life, you have to be patient with it. You can't push it to perform but if you give it time, it will."

Garden Gallery

Hollywood Dazzler

Nestled in the hills above Hollywood, California, sits Villa Pina. Its five small garden terraces step their way up the steep hillside. Their gardens are a colorist's dream and a designer's challenge. Landscape designer Laura Morton took on Villa Pina's dream and challenge at the request of its owners. For them, the 1930s Spanish Revival architecture and its hillside location were reminders of their roots—his in Capri, hers in Spain. They asked Morton to design a garden that would enhance the home's European feel and offer color, color, and more color. They also wanted each terrace to be its own distinctive world.

Morton spent much of her late teens and early twenties working as a fashion model in Europe. From working in different countries and with so many different creative disciplines, she developed a unique

Landscape designer Laura Morton designed a multilevel garden for Villa Pina in Los Angeles's Hollywood Hills. Each level is its own garden terrace with its own European-influenced theme and color scheme.

Morton designed this fountain after grottos she'd seen in Europe. She incorporated salvaged red roof tiles, set upside down for the water to splash into.

style that she uses to pull together concepts for new spaces. When it came to creating a space with an Old World feel, she knew just how to approach it.

When Morton first inspected the property, she found it "an overgrown spider and lizard infested jungle that hadn't been managed in a long time," she says. Retaining walls were covered with a thick layer of old *Ficus repens* (creeping fig). *Tecoma capensis* (Cape honeysuckle) had taken root everywhere, jumping from one terrace to another. The honeysuckle and creeping fig had both worked their way up the trees. A ficus hedge had been planted to hide failed retaining walls, while access to the top pool terrace and its city views required a terrifying trip up a narrow stairway. The old spray-head irrigation system had left the shady surfaces slimy and dangerous.

As demolition began, it was clear that most of the walls needed rebuilding or reinforcement. Ancient paint was sandblasted away and walls were coated with fresh, breathable plaster. Morton began the new garden's color transformation by

The Den of Moors, the lowest terrace, has deep yellow ochre walls with persimmon details, and this raised fire bowl as its focal point. Above it, the Jar Terrace is named for the tall olive jar focal point. Ironically, the jar hides an old olive tree stump that proved too difficult to extricate from the high, narrow terrace.

When retaining walls were being redone, Morton used natural earth pigments applied "al fresco," before the plaster was fully cured. Layers of color like this traditional ochre create the rich, burnished European look the designer and homeowners were after.

OPPOSITE TOP: Unthirsty plants for this home's climate repeat and offer contrast to the ochre walls, including *Rosa* 'Chocolate Sundae' and succulent lavender *Echeveria* 'Afterglow'.

OPPOSITE BOTTOM: Outside the owners' bedroom, Morton created a small, romantic park that includes a square of grass maintained with a high-efficiency, belowground irrigation system.

Garden Gallery

adding pigments to the walls of each terrace. To get that old European look, she drew upon tradition. Rather than use paint, Morton chose natural earth pigments applied "al fresco," meaning that they were applied before the plaster was fully cured. Layers of color created the rich, burnished look Morton was after.

Each terrace has its own color scheme and its own plant palette, yet with Morton's keen design eye, the connection from one to the next is unmistakable. At the lowest level, the home's living room doors open to a sumptuous outdoor room, known as the "Den of Moors" for its Moroccan influence. The room has deep yellow ochre walls with persimmon-colored details. A poured concrete daybed is topped in richly upholstered persimmon cushions. The daybed backs onto a teak Indian *jali*, an antique carved screen. The room's focal point is a raised fire bowl modeled after the octagonal fountains typical of Moorish styling.

Dusky red *Rosa* 'Chocolate Sundae' (climbing rose) and golden-flowered *Lonicera confusa* (honeysuckle) climb the ochre walls, up toward what Morton calls the "Jar Terrace," named for a 4-foot-tall olive jar that serves as that space's focal point. Ironically, the jar stands on an old olive tree stump that proved too difficult to extricate from the high, narrow terrace.

The Jar Terrace and the Den of Moors below feature unthirsty plants that repeat and contrast the ochre walls: deep burgundy–leaved *Cotinus coggygria* (smoke tree); green and yellow variegated agave; red-, yellow-, and green-striped *Leucadendron* 'Jester' (conebush); bright red–flowering geraniums; burgundy- or black-leaved succulent *Aeonium* 'Zwartkop'; red-barked manzanita; and a brilliant magenta

Morton tucked an upright Italian cypress into a corner along with the tall *Neomarica caerulea* (walking iris) and purple-flowering *Alstroemeria* (Peruvian lily), along with *Pelargonium* 'Veronica Contreras' to bloom bright purple-pink against the violet background.

BELOW: *Rosa* 'Veilchenblau' is stunning against violet-pigmented walls on the park terrace.

The Birdbath Terrace is largely planted with California natives whose nectar and berries attract birds to this very urban setting. It is a study in color contrasts: the bright blue birdbath along with lime green, silver, and variegated green and chartreuse foliage all stand out against deep terra cotta–colored walls.

purple-flowering *Bougainvillea brasiliensis*, a favorite of the homeowners.

Above the Jar Terrace and outside the owners' bedroom, Morton created a small, romantic "park." There is a small square of grass irrigated with high-efficiency in-line drip irrigation installed several inches below the surface. Walls are colored with a traditional pigment known as "Colonial Violet." Here, Morton went tone-on-tone with plants such as plum-leaved *Cercis canadensis* 'Forest Pansy' (eastern redbud) and *Vitis* 'Roger's Red' grape, whose green leaves turn burgundy in fall.

Mounds of *Pelargonium* 'Veronica Contreras' (pansy-faced geranium) bloom bright purple-pink against the dusky violet background. In spring, a rambling *Rosa* 'Veilchenblau' trained against the wall erupts in fragrant, bright violet flowers with white centers. A simple bench completes the parklike effect.

Further along, the Birdbath Terrace features a vintage birdbath as its focal point and color contrast. Morton painted the birdbath robin's egg blue, then surrounded it with mostly native plants whose nectar and berries attract birds. Pink-flowering

Garden Gallery

The Fountain Terrace is the home's gathering place and outdoor dining room. Foliage from *Wisteria* 'Black Dragon' covers a bent metal arbor to create a leafy ceiling. Fiery red pomegranate flowers bloom in the corner.

Salvia spathacea (hummingbird sage), *S. apiana* (white sage), and native grasses form the "carpet" in this room. *Heteromeles arbutifolia* 'Davis Gold' (toyon) is native but uncommon in that its berries are gold rather than red.

Chartreuse and deep green variegated *Ceanothus griseus* 'Diamond Heights' (California lilac) is espaliered on one terra cotta–colored wall. Exotic *Acacia boormanii* (Snowy River wattle) along with a fruiting *Acca sellowiana* (pineapple guava) all survive on the same spare irrigation schedule. Throughout the garden, irrigation is controlled by an irrigation controller that uses satellite data to determine exactly when and how much or how little to irrigate.

The Fountain Terrace serves as Villa Pina's outdoor dining room. Its walls are a burnished pomegranate red, using a pigment called "Plum." Here, perhaps more than anywhere else in the garden, one see's Morton's mastery of color and media. "I like the idea of layering and enhancing the colors," she says, "by pulling out the various tonalities of the natural pigments. Sometimes you do that with contrast."

An antique Indian door picks up the room's red tones and adds contrasting, aged verdigris.

In this garden room, a pomegranate tree with its flame orange flowers stands out against the deeper red walls. The rich, red mahogany trunks and branches of *Arbutus* trees, planted from 36-inch boxes, look like sculpture against the walls. The multitone gold-to-cerise flowers of climbing *Rosa* 'Joseph's Coat' pull out all the reds and oranges in terra cotta, adding incredible vibrancy.

To form the floor of this room, reddish granite cobble pavers are laid in a fish-scale pattern and set in sand so water percolates through. Morton designed a bent metal arch over the table and planted it with an unusual, deep purple–flowering *Wisteria* 'Black Dragon'. Mock candles illuminate the night, creating an inviting space for a glass of wine on a summer's eve.

One corner of the room is filled with the fountain for which the terrace is named. Morton envisioned a grotto as she created the design. She incorporated salvaged red roof tiles, set upside down so water would splash off them. An antique Indian door at the rear of the terrace picks up the red tones and adds contrasting verdigris (a greenish blue patina). The door suggests the magic and mystery of another space to explore just beyond. (Alas, it's all a trick: behind the door is bare hillside rather than another garden room.)

When asked about her use of color, Morton says that rather than taking a scientific approach, "it's more instinctual. I'll try a combination and if it works well, I want to work with it more and more…. like mixing violet blue with pomegranate red. It works so well with orange, violet, pomegranate red, and black—like black pots and black ironwork…. it's magic."

FACTS AND FIGURES

ELEVATION 508 feet
ANNUAL AVERAGE RAINFALL 18 inches, between November and March; dry in summer
ANNUAL RAINY DAYS 24
SUMMER HIGH 85–86°F, typically July to September
WINTER LOW 45°F in December and January
HUMIDITY Ranges from 87–89% in May, July, and September to 60% in December
IRRIGATION All in-line drip irrigation; in the lawn area, the drip is buried several inches deep
SOIL TYPE Varies throughout the property from clay to decomposed granite to loam
DOMINANT PLANT TYPE Mixed Mediterranean
USDA ZONE 10b

KEY PLANTS FOR COLOR

Aeonium 'Zwartkop'
Arbutus 'Marina'
Bougainvillea brasiliensis
Cotinus coggygria
Echeveria 'Afterglow'
Kalanchoe luciae
Leucadendron 'Jester'
Salvia clevelandii 'Winnifred Gilman'
Salvia spathacea
Vitis 'Roger's Red'

Garden Gallery

Colors on the Hill

As you head north out of Santa Barbara, California, you drive through a tall mountain pass whose finger canyons weave back into oak forest and away from traffic. Follow one of those roads to find the home of Jay Bassage and Shelly Vinatieri. Here, on warm spring and summer afternoons, the fog rolls in from the nearby ocean and blankets their 40 acres in a thick layer of white. When it recedes, the couple's beautiful home and garden slowly emerge from the fog, as if by magic.

This secluded property was once a commune; today, it is a modern craftsman home designed by Bassage. The garden is the creation of San Luis Obispo–area

The homeowner designed this modern craftsman home on 40 hilly acres in Santa Barbara, California. The garden was created by San Luis Obispo–area landscape designer Nick Wilkinson.

Garden Gallery

Red, green, yellow, orange, and soft blue dominate the landscape. Plants are arranged to create dramatic contrasts of height, volume, texture, and color.

OPPOSITE: The scale is vast but the feeling intimate thanks to this small flagstone patio surrounded by succulents and broad-leaved plants that direct the eye to the mountains and native chaparral that surround the home.

designer Nick Wilkinson. Wilkinson is a painter, sculptor, expert plantsman, and owner of both Grow, a quirky succulent boutique in Cambria, California, and Left Field art gallery in San Luis Obispo.

On a visit to Cambria, Bassage and Vinatieri stayed at an inn at nearby Moonstone Beach where they fell in love with the garden Wilkinson designed. The colors, the texture, and Wilkinson's light hand spoke to them. "The garden wasn't crowded," Bassage says, "Nick allowed for blank spaces on the canvas."

Wilkinson used the same light hand to design the couple's garden. Their hillside home has taupe-colored walls accented by burgundy trim, all set against a deep green background of California native oaks. The garden's framework was already set. Arizona flagstone–topped retaining walls form a series of tiers. The artist in Wilkinson treated each tier as a gallery and placed plants as if they were sculptures.

Red, green, yellow, orange, and soft blue are the dominant colors in this melding of succulents, cacti, and dry-climate, broad-leaved plants. Against the deep neutral palette of the home and hardscape, the bright colors bring the garden to life.

At house level, plants are soft, leaves are round, and the feeling is inviting. Ice green–leaved manzanita, bright burgundy–leaved

Reds repeat throughout the garden: in the home's trim, in the red foliage of *Euphorbia cotinifolia*, in the red blades of succulent *Kalanchoe luciae*, in the burnished tips of *Aeonium* rosettes, and of course in the stunning flowers of perennial *Anigozanthos* 'Big Red', a native of Australia.

At house level, soft plants and round leaves contribute to the inviting feeling.

Euphorbia cotinifolia (Caribbean copper plant), and tricolored *Leucadendron* 'Jester' (conebush) are the foundation plantings that bridge the surrounding forest of live oaks with the planted landscape.

From the lower driveway, the uphill view toward the house enhances the garden-as-gallery effect. First, Wilkinson mounded soil in tiers to create dimensions that Bassage fondly describes as "rolling hardscape." Then, in Wilkinson's hands, plants are arranged to create dramatic contrasts of height, volume, texture, and color, using the bright orange blades of *Euphorbia tirucalli* 'Sticks on Fire' to set the garden's tone. Wilkinson added a wide selection of agaves in soft yellow and green variegation, blue, and deep green—some that were artichoke shaped, some soft bladed, rigid bladed, and so on.

Smaller, shrublike, and spiny *Euphorbia milii* (crown of thorns) has round, deep green leaves and bright red flowers. Low-growing *Senecio mandraliscae* (blue chalk fingers) snakes through the beds, around almost-black *Aeonium* 'Zwartkop', bright red, orange, and yellow flowering kangaroo paws, and lavender-leaved *Aloe* 'Blue Elf', whose ever-present, bright orange flowers match *Euphorbia tirucalli* 'Sticks on Fire'.

The side garden features another of Wilkinson's succulent compositions, this one surrounding a large, east-facing flagstone patio and a water-thrifty buffalo grass (*Buchloe dactyloides* 'UC Verde') lawn. The lawn serves as the garden's negative space, as well as a play space for dogs and their humans.

In the surrounding garden beds, Wilkinson used colorful broad-leaved plants like *Leucospermum* (flowering

Euphorbia cotinifolia (right) has burgundy leaves that match the home's trim. Its round leaves and rounded tufts of foliage contrast with the surrounding plantings and meld into a deep green background of native live oaks.

pincushion), blue-gray-leaved *Cussonia paniculata* (cabbage tree), burgundy- and green-leaved *Brachychiton rupestris* (bottle tree), and *Beaucarnea guatemalensis* (ponytail palm). Large succulent agaves add shape, while large aloes contribute

ABOVE: After living with Wilkinson's garden creation, Bassage and Vinatieri have an entirely new appreciation for the possibilities of low-water gardens.

OPPOSITE TOP LEFT: Wilkinson set the color palette in part by using *Leucadendron* 'Jester', a drought-tolerant South African native. Its red, yellow, and green tricolor blades repeat throughout the plantings and are a major player in the garden's year-round color.

OPPOSITE TOP RIGHT: Madagascar native *Alluaudia procera* (Madagascar ocotillo) has upright branches covered in spines and round green leaves, which together exemplify the way Wilkinson uses plants as sculpture. As it has grown, a branch settled onto the wall's sandstone cap, suggesting an elbow at rest.

OPPOSITE BOTTOM: Wilkinson's love of unusual plants plays out in the gardens he designs. This young planting, seen from above, includes orange-blooming *Anigozanthos* 'Bush Tango' and yellow-blooming 'Bush Dawn', along with a tall tree aloe, *Beaucarnea guatemalensis* (ponytail palm), yellow- and green-striped *Agave* 'Joe Hoak', and the swirl of emerging gray leaves of *Cussonia paniculata*.

color in winter with their upright coral candelabras. This is one of the couple's favorite spots in the garden. Bassage says, "It is wonderful in the morning to sit in the den and look at the garden in the morning light."

Wilkinson worked with landscape contractor Gabriel Frank and his crew for months on the project. Bassage jumped in, moving in large boulders and small rocks from other areas of the property. Finally, the ornamental beds were mulched with a thick layer of gravel called "Ice Sand," whose color matches the hardscape

Under Wilkinson's tutelage, the couple has learned much from their garden. Bassage says, "I want people to know that the availability of variety of drought tolerant plants is incredible. The colors and variegations are infinite so anyone can create their own palette."

FACTS AND FIGURES

ELEVATION 1392 feet
ANNUAL AVERAGE RAINFALL 20 inches between November and March; dry in summer
ANNUAL RAINY DAYS 26
SUMMER HIGH 81°F, typically in August
WINTER LOW 38.9°F in January
HUMIDITY Ranges from about 90% in May, July, September, and October to 60% in January
IRRIGATION In-line drip irrigation except on the lawn, which is sprayed with overhead spray
SOIL TYPE Sandy loam with a bit of clay
USDA ZONE 10b
DOMINANT PLANT TYPE Succulents with some broad-leaved Mediterranean climate plants

KEY PLANTS FOR COLOR

Aeonium 'Sunburst' and 'Zwartkop'

Agave 'Joe Hoak'

Agave vilmoriniana 'Stained Glass'

Aloe 'Blue Elf'

Anigozanthos 'Big Red' and 'Bush Dawn'

Crassula ovata 'Hummel's Sunset'

Euphorbia cotinifolia

Euphorbia tirucalli 'Sticks on Fire'

Kalanchoe luciae

Leucadendron 'Jester'

Senecio mandraliscae

Cool Colors

If you've ever visited the Hollywood sign, one of Los Angeles's most iconic landmarks, you've likely driven past Judy Horton's garden. Not that you'd know it; decades ago, someone planted the front of her property with a ficus hedge. Today it's so tall that the only glimpse you'll catch of her garden is through a small arch cut into the hedge for a wrought iron gate. What you'd see in that moment's view, however,

Located near the base of the Hollywood sign, landscape designer Judy Horton's home is one of the original 1920s Hollywoodland cottages. The unusual combination of the minty sage stucco with Moroccan blue tile inspired her color scheme of intense blue, turquoise, rosy pink, and yellow similar to the garden of French artist Jacques Majorelle in Marrakech. Here, a pale yellow-blooming *Grevillea* 'Moonlight' makes a statement at the front entry, opposite an unusual variegated *Ligustrum sinense* 'Variegatum' (Chinese privet).

Garden Gallery

Red-leaved *Leucadendron* 'Safari Sunset' (conebush) stands out against shiny Moroccan blue roof tiles.

The home's original landscape was just lawn with a few small flowerbeds. Today it is a Mediterranean garden that features the many colors of Mediterranean and Californian plants.

is a classic Spanish bungalow, one of the original Hollywoodland cottages from the 1920s. When Horton first came across the property, she was searching for a traditional casita with white stucco and a red tile roof. Instead, this little house was stuccoed on the mint side of sage green with a blue Moroccan tile roof, an unexpected color scheme to say the least. Still, the landlord was great and the location was perfect. "I didn't think I'd stay here more than two years," Horton recalls, "I was desperate to have a garden right away so I thought, 'Okay, I'll go with the blue tile.'"

Fortunately for Horton, she had designed a client's garden with a similar color scheme not long before. The palette of intense blue, turquoise, rosy pink, and yellow was borrowed from the French

Every inch of this small urban lot counts as garden, including this narrow side yard, home to one of Horton's potted plant collections.

Horton reduced the scale of a massive ficus hedge by adding a low, curved hedge of small, silver-leaved *Teucrium fruticans*. A dog-secure chain-link fence is sandwiched between the two hedges.

The narrow side yard outside Horton's office includes a fence painted Majorelle blue, adorned with delicate-looking plants such as pink-flowered *Pelargonium* 'Gary's Nebula', bronze-speckled *Billbergia* bromeliads, orange- to pink-flowered *Bougainvillea* 'Tahitian Dawn', and golden-flowered *Bougainvillea* 'California Gold'.

Vitis 'Roger's Red' shows off a red-on-blue scheme in fall.

artist Jacques Majorelle's garden in Marrakech, Morocco.

Besides the oversized hedge, the home's original landscape was basic lawn with a few small flowerbeds. From the main living area, the view was of the hedge and a worn concrete drive leading to an old-fashioned single-car garage.

Horton arrived with dogs, furniture, and a big collection of potted plants, some dug up from her former home. "I already wanted a garden that was basically Mediterranean using the colors of the Mediterranean and California plants," she says.

Horton replaced narrow flowerbeds with dry-growing, succulent and nonsucculent plants. This tone-on-tone combination includes *Agave* 'Blue Glow', *Origanum dictamnus*, *Senecio mandraliscae*, and *Aloe plicatilis*.

With her landlord's blessing, Horton removed more than eighty percent of the lawn. Next, she planted a curved *Teucrium fruticans* (bush germander) hedge just inside the ficus hedge. She chose bush germander for its small, silver leaves and blue flowers that contrast the dark green ficus both in terms of color and texture. At more than 4 feet tall, the clipped bush germander softens the ficus hedge's angularity and brings down its imposing scale. At the same time, the smaller hedge is tall enough to hide a chain-link fence that keeps Horton's dogs in the inner yard.

"Some decisions are practical while others are inspired by architecture," Horton says, "or, one plant leads to another and you suddenly have a wonderful color scheme going."

To block the view of the driveway, Horton layered in more tall plants, including native *Heteromeles arbutifolia* (toyon). Back in the 1920s, toyon covered the local hillsides. The evergreen shrub's bright red winter berries reminded area developers of holly berries, which led to the name "Hollywoodland." In Horton's garden, the

Garden Gallery

berries add a red color pop in winter and serve as tasty snacks for native birds.

Horton's bedroom and office look into the opposite side of the 5,000-square-foot property. Here, a narrow strip of sidewalk and a tall fence separate her garden from her neighbor's. Since Horton spends much of her day in these two rooms, she wanted to see a garden through the windows. She painted the fence Majorelle blue, then chose a different combination of plants, in part because this small, protected area accomodates more delicate and thirsty plants. A pot of coral pink–flowered *Pelargonium* 'Gary's Nebula' (geranium) is stunning against the blue wall, as are a pair of bougainvilleas: 'Tahitian Dawn' opens with orange flowers that mature pink, while 'California Gold' erupts in cascades of golden bracts. In fall, deep red–leaved *Vitis* 'Roger's Red' picks up the red-on-blue scheme. Horton prunes the vine to delicately trace her bedroom and office windows.

Back in the main garden, Horton replaced narrow flowerbeds with a combination of succulent and nonsucculent plants, all dry growing. There's a texture-focused tone-on-tone combination that includes red-edged *Agave* 'Blue Glow', the felty silver leaves of *Origanum dictamnus* (dittany of Crete), the narrow, succulent blades of *Senecio mandraliscae* (blue chalk fingers), and *Aloe plicatilis* (fan aloe), whose long tongue-shaped leaves reveal a yellow edge when the sun hits them just right. In early spring, orange—from aloe flowers, native California poppies, dry-growing *Clivia*, and a large, potted kumquat tree that Horton keeps on her front porch—offers contrast to the greens and blue greens.

Studies in tone-on-tone.

Garden Gallery

Horton's design surrounds her house with gardens so every window has a view.

There are other color themes throughout the garden. In late spring, native *Eriogonum giganteum* (St. Catherine's lace), toyon, and *Lychnis coronaria* (rose campion) all bloom white to cream, picking up on the pale yellow–blooming *Grevillea* 'Moonlight', and the cream and green variegated *Ligustrum sinense* 'Variegatum'.

Horton's rear patio is where the thirstier plants are concentrated. Three of its walls are the same minty sage green, while the fourth is bright Majorelle blue, complemented with blue-blooming hydrangeas in very large containers. Since hydrangeas are thirstier than most other plants in the garden, growing them in containers allows Horton to adjust her watering to keep them going year-round. She adds just enough aluminum sulfate to keep them blue.

Horton also favors deep burgundy leaves, especially when they contrast with green and yellow variegated leaves. She has a pair of enormous copper beech trees, for example, that she purchased before learning that they wouldn't grow in Southern California. Luckily, she had not yet planted them in the ground. Instead, they went into huge pots where they are watered regularly, fertilized annually, and root pruned every three or four years.

The beech trees create a little courtyard just outside Horton's garage. "Then, one day I brought home a little brown flax called 'Chocolate Baby'," Horton says, "and a *Pittosporum tenuifolium* called 'Silver Magic', which is a reverse variegation from 'Silver Sheen'." The combination would never work in the ground, she explains, because the pittosporum would get too big. But in pots, she can control the sizes so the plants work well together. After that, Horton added plants with chartreuse foliage or yellow variegation and more burgundy foliage, all to test with the beech trees. "I did it for years," she says, "then used that color scheme in a client's garden."

Horton's irrigation is an antiquated overhead spray system supplemented by bubblers. Since she rents the property, there are some limitations to her investment. Water drains right through her sandy soil. As a result, she has to water once a week most of the year. In between, if she encounters a plant looking a little droopy, she gives it a squirt from one of the watering cans stashed throughout the garden. Still, Horton's water use has never exceeded the lowest price tier charged by the City of Los Angeles.

Horton is most pleased with the way the bones of the garden support the overall presentation. "When I start a design, I look for the "big idea" or concept," she says. "Once I find that, the rest is detail. The big idea for my own garden was surrounding the house with gardens so that every window has a view, if only of a Buddha's hand citrus in a pot, or a glimpse of grape foliage."

FACTS AND FIGURES

ELEVATION 667 feet
ANNUAL AVERAGE RAINFALL 18.3 inches, mostly falling from November to March; dry in summer
ANNUAL RAINY DAYS 24
SUMMER HIGH 76°F, typically in August
WINTER LOW 55°F in December
HUMIDITY Ranges from 87–89% in May, July, and September to 60% in December
IRRIGATION Ancient overhead spray, bubblers, hand watering
SOIL TYPE Sand
USDA ZONE 10b
DOMINANT PLANT TYPE Mixed Mediterranean

KEY PLANTS FOR COLOR

Ceanothus 'Concha'
Clivia nobilis and *Clivia* hybrids
Cotinus coggygria 'Royal Purple'
Eschscholzia californica
Punica granatum
Salvia clevelandii 'Winnifred Gilman'
Sedum nussbaumerianum
Senecio mandraliscae

Garden Gallery

Colorful Subtlety

Deserts are widely appreciated for their subtlety, so it comes as no surprise that desert gardens follow suit. These gardens, especially the low-water, high-desert gardens of New Mexico, seldom include plants that scream out their colors. Instead, colors are subtle, offered in small doses, and show up best when set in contrast to other colored elements. Such is the case with Doreen and Phillip Radcliff's Albuquerque garden.

Doreen Radcliff is the gardener in the family and is obsessed with color. Inside, the walls of the couple's home are lined,

Colors In New Mexico's high-desert gardens are subtle and come in small doses. This garden, designed by naturalist and author Judith Phillips, uses belowground drip irrigation and is mulched with rock and crusher fines. In the heat of summer, the garden is irrigated for three hours once every ten days at most, to encourage deep, drought-protected roots.

Garden Gallery

Phillips uses a diversity of plants to support birds and beneficial insects. Here she combined yellow, daisylike *Berlandiera lyrata* (chocolate flower) with strappy *Nolina texana* (bear grass) and soft purple *Aethionema schistosum* (Persian rockcress).

literally, with bright-colored paintings of green-throated hummingbirds, red robins, blue jays, yellow sunflowers, and desert scenes framed in gold, turquoise, magenta, and lime. Doreen's penchant for bright décor extends to the garden. "Color is cheerful," she says, "it brightens the world and makes everything so much better."

Because the Radcliffs' home sits at the crook of a cul-de-sac, the wedge-shaped garden is small, but its impact is big. Noted New Mexico designer, author, and naturalist Judith Phillips designed the garden in 1998. Doreen Radcliff asked Phillips for a garden that required minimal water while still offering lots of flowers and fragrance. As a very hands-on gardener, Radcliff intended to tinker with the plantings over time, and wanted diversity in the garden to attract wildlife.

Diversity in plantings is important to Phillips, too. "Ecologically it is a better idea for birds and insects, beneficial insects," she says, and adds that mass plantings of just a few types of plants are also more susceptible to pest problems.

From a design standpoint, Phillips explains that it takes more than plants to create a cohesive garden, especially in a climate where seasonality translates to plants with highs and lows. "Plants reinforce the coherency," Phillips says, "but it's the hardscape that makes sense year-round." To that point, the Radcliffs' front garden is divided into two distinct spaces. The streetside plantings wrap around a wall that separates the public side of the garden from the private inner courtyard garden. Entry to the courtyard is through a decorative, rust-colored steel gate that Doreen Radcliff designed and had fabricated. Plants are more protected inside the courtyard. That gives the space two distinct garden zones, both in terms of plant hardiness and in the ability to incorporate *objet d'arte* into the garden.

Phillips always designs with site and climate in mind. Her original design for the Radcliff home featured *Forestiera neomexicana* (New Mexico olive) as a canopy tree to shade the home and garden. A few years later, however, a fire caused serious damage to the property. Much of the front garden had to be replanted, and the New Mexico olive never recovered. Phillips replaced it with *Quercus fusiformis* (escarpment live oak), which stays evergreen in Albuquerque and grows 20–25 feet tall by 12 feet wide. The rest of the

This front yard garden is divided into streetside plantings set around a privacy wall and an inner courtyard garden. Homeowner Doreen Radcliff designed the gate.

Garden Gallery

Even the family cat admires the hot pink-flowering *Cistus* ×*purpureus*.

Daisylike chocolate flower (*Berlandiera lyrata*) blooms through most of the growing season. As its name would suggest, it really does smell like chocolate.

OPPOSITE: Though the garden sleeps from late November through March, it is planted so that blooms arrive in early spring, then continue in waves of purple, pink, yellow, and coral until fall.

garden is filled with woody shrubs, some vines, and many perennials that Doreen Radcliff adds to and updates over time.

Though the garden sleeps from late November through March, it is planted to ensure that blooms start as early as possible in spring, then continue in waves of purple, pink, yellow, and coral until fall's chill.

The earliest blooms include hot pink- and bright white-flowering species of *Cistus* (rockrose), along with chartreuse-flowering *Euphorbia myrsinites* (myrtle spurge) and clusters of the fragrant, pale lavender flowers of *Aethionema schistosum* (Persian rockcress).

In summer, *Bignonia capreolata* 'Tangerine Beauty' (crossvine) joins the bloom. Its big, trumpet-shaped, coral orange flowers are offset by cheery clusters of golden yellow *Zinnia grandiflora* (desert zinnia). Those bold colors and textures are

Garden Gallery

Plants in the courtyard play off accessories and artwork: a lavender gazing ball sits among the gray-green foliage of *Salvia lavandulifolia* (left), while a copper oil lamp sculpture rests atop a rustic stump, offering contrast to the pigmented stucco wall (right).

balanced by the *Salvia dorrii* (desert sage), whose clouds of tiny, deep purple flowers are set against soft green–colored leaves.

By fall, the garden is filled with billowy clouds of purple and pink from *Muhlenbergia capillaris* 'Regal Mist' (pink muhly grass) and wands of bright purple–dotted *Liatris punctata* (gayfeather). Fragrant, butter yellow, daisylike *Berlandiera lyrata* (chocolate flower) blooms through most of the growing season.

Inside the courtyard, plants play off accessories and artwork. A deep lavender gazing ball is nestled among the gray-green foliage of *Salvia lavandulifolia* (lavender sage). A copper oil lamp sculpture by artist Chris Caldwell is set atop a rustic stump; its verdigris patina contrasts against the stucco wall, whose soft tan-green pigment is called "Cottonwood."

Winter lows outside are often in the 20s and sometimes lower, but inside the courtyard, it stays about five degrees warmer. Plants like the purple-flowering *Leucophyllum langmaniae* 'Lynn's Legacy' (Texas sage), tend to stay evergreen and bloom longer.

The conditions in this garden, Phillips says, are typical for the region. The decomposed granite soil is extremely compacted, so the soil gets loosened only in the planting holes, and only as they are dug. No amendments are added, no rototilling, and no fertilizer. "The plants don't need fertilizer," Phillips explains. "Arid-climate plants have a very low nutrient demand. Microorganisms in the soil associate with the plants to help them assimilate what they need (from the soil)." If you fertilize, Phillips continues, you can actually destroy the mycorrhizae, which are fungal filaments that move water and nutrients through the soil to plant roots. Even worse, fertilized plants grow "like crazy" and become more susceptible to insects. "Its a ripple effect," she says, "and none of it is good."

Phillips and Radcliff are both seriously conservative when it comes to water. The entire garden is mulched, some with rock and some with crusher fines, the dust and slivers of rock leftover from crushing rock to make gravel. The drip irrigation system's flexible polyethylene lines connect to a PVC skeleton underground. Narrow poly tubes snake up to the soil surface, and connect to individual drip emitters by each plant. Emitters typically release just one or two gallons per hour. There are only as many emitters as each plant requires.

The ever hands-on Radcliff manages her irrigation system rather than relying on an automatic controller. From May through August, she irrigates once every ten days at most, and then for three hours at a time.

Doreen Radcliff's garden demonstrates how to have a beautiful and waterwise garden by choosing the right plants, such as purple-flowering *Salvia dorrii*.

Those long waterings ensure that water goes deep, which initially helps plants grow deep, drought-protected roots, and later ensures that water reaches those roots. "After 25 years of gardening in New Mexico," Radcliff says, "I see people using their drip irrigation incorrectly. They water for ten minutes every other day. When you do that, you encourage roots at the soil surface. Those roots burn. What you want is for roots to go deep, that's why you water longer. You can space out irrigations longer because it's still semi-damp deep, where the roots are."

When summer monsoon rains arrive at the end of June, Radcliff skips irrigation cycles, or extends the intervals between cycles rather than waste water on plants when they don't need it. In September, waterings are spaced out to once per month and always for the same duration. In winter, if it snows, the garden isn't watered at all. "You really can have a beautiful garden with very minimal water usage if you choose the right plants," Radcliff continues. "When you think about it, it's common sense."

FACTS AND FIGURES

ELEVATION 5,640 feet
ANNUAL AVERAGE RAINFALL 10.3 inches, falling mostly in July and August
ANNUAL RAINY DAYS 28
SUMMER HIGH 100°F in July
WINTER LOW 23°F in December and January
HUMIDITY Ranges from 92% in August to 60% in March and October
IRRIGATION Drip, with individual emitters to plants
SOIL TYPE Decomposed granite
USDA ZONE 7b
DOMINANT PLANT TYPE Desert broadleaf

KEY PLANTS FOR COLOR

Berlandiera lyrata
Bignonia capreolata 'Tangerine Beauty'
Cistus species
Leucophyllum langmaniae 'Lynn's Legacy'
Liatris punctata
Penstemon ambiguus
Salvia chamaedryoides
Salvia dorrii
Solidago spathulata var. *nana*
Zinnia grandiflora

Garden Gallery

Seaside Color

Cayucos is a tiny town on California's central coast, named for the small, canoelike fishing boats used by the native Aleut people to hunt sea otters. About 2,500 people call Cayucos home, including Judy Mae Straw and her husband, Reg Whibley. The couple owns a 60-year-old beachfront cottage. While beach cottages are often cute, theirs is more than cute; the cottage and the garden that surround them are eye candy.

Large glass doors focus the view from inside out to the waves just a few hundred feet beyond what the couple considers their front yard. Where the ocean is moody, the hardscape and plants are

The tiny beach town of Cayucos, California, is home to this tidy, amazing garden whose vivid colors and varied textures look beautiful, even under moody skies. The garden was created as collaboration between the owners, their friends, and artist and landscape designer Nick Wilkinson.

bright and cheery, thanks to vivid colors and varied textures. A wide wooden deck and broad, shallow steps wrap around the cottage to serve as its patio. Master carpenter Whibley built the deck, while a friend tiled the step risers in red, yellow, blue, green, black, and white—all arranged in graphic patterns inspired by nautical motifs.

At the next level down, Whibley designed a grotto and patio of undulating curves. It's a crazy quilt of rock, brick, broken glass, and "found" objects. Straw's late mother's collection of miniature perfume bottles fill holes in some of the bricks; old metal bottle caps fill others; and blue glass pieces from a shattered windshield fit in gaps between rocks. A river of turquoise, teal, blue, and white glass pieces flows through the hardscape, terminating at a stone fireplace. These design elements swirl and tumble in waves like the ocean.

The hardscape transitions to a garden of brilliant red, green, and yellow—mostly succulents—designed by artist and landscape designer Nick Wilkinson. Wilkinson built upon the hardscape that the couple had already created. On his first visit to the garden, Wilkinson says, "it was a painting that was half done, but already looked like a masterpiece." It's a small space, only 30 by 30 feet, yet you can get lost for hours just looking at the detail, appreciating the craftsmanship, and finding new, exciting elements.

Wilkinson arranged the plantings in a semicircle to frame the space. Yellow-, pink-, and green-leaved *Aeonium* 'Sunburst' and black-leaved *A.* 'Zwartkop', variegated *Agave vilmoriniana* 'Stained Glass' (variegated octopus agave), orange-blooming *Aloe vanbalenii*, and orange-leaved

This deck, built by Whibley, wraps around the cottage and serves as its patio; a friend tiled the step risers in patterns inspired by nautical motifs.

A grotto of undulating curves—made up of rock, brick, broken glass, and "found" objects such as miniature perfume bottles, old metal bottle caps, and pieces of glass from a shattered windshield—sits just below the deck.

ABOVE: Wilkinson's artistry and plantsmanship are evident in this pairing of the icy green blades and deep rose flower bracts of *Euphorbia rigida* (gopher spurge) with a frilly, red edged succulent, *Echeveria* 'Mauna Loa', and the flowers of gold-edged *Aloe distans* in the background.

LEFT: Red-edged *Echeveria agavoides* 'Lipstick' shines in a cobalt blue pot.

OPPOSITE: Wilkinson mass planted rosettes of the red-flowering succulent *Aeonium nobile* against a background of bright green, upright *Euphorbia mauritanica* (pencil milk bush). He tucked flat-bladed *Aloe plicatilis* (fan aloe) among the rocks, and added green and cream *Euphorbia characias* (variegated spurge).

Garden Gallery

Crassula capitella 'Campfire' combined with the colorful, rosette-shaped *Echeveria* 'Etna' and 'Mauna Loa', and dusty blue *Senecio mandraliscae* (blue chalk fingers) are just a few of the many textural succulents in this garden, all chosen for their mounding shapes and bright colors. "The garden was designed to capture the shapes of the ocean…. that makes it appropriate for the space," says Wilkinson. "I wanted it to be fun, interesting and as quirky as the people who inhabit the house," he adds with a smile.

Wilkinson's choice of low-growing plants ensures an unimpeded view of the ocean. Straw added ceramic pots in intense cobalt blue, burnished green, terra cotta, and bright yellow, all colors found in the landscape. She fills them with brightly colored succulents to tie the softscape to the wood deck.

From house level, the overhead view of the garden looks like a painting with intense colors that stand out against the dove gray sand, pale blue-green water, and often gray skies.

There is a drip irrigation system but the garden is watered only occasionally, despite constant seaside winds. It's an undemanding garden that Straw happily cares for; "I love getting my fingers dirty," she says.

Straw and Whibley are eager to share the whimsical space they've created. "There's a bench along the public walkway down to the beach," Straw says. "I put it there for people to sit on and enjoy the garden. I'm thrilled when they use it."

ABOVE: Rounded rocks, plants, and landforms roll and tumble like the waves of the ocean.

OPPOSITE TOP: This color-filled, low-growing plant tapestry allows unimpeded views from house to ocean. The garden includes the curved arms of *Aloe vanbalenii*, ghostly white, native *Dudleya pulverulenta* (chalk liveforever) in a "lawn" of *Senecio mandraliscae* (blue chalk fingers), the red-burnished blades and flowers of *Aloe distans*, with its gold tooth edges, green and yellow variegated *Agave attenuata* 'Kara's Stripes', and *Euphorbia characias* 'Tasmanian Tiger', all topped with a young *Dasylirion longissimum* (Mexican grass tree).

OPPOSITE BOTTOM: Straw added ceramic pots in a variety of colors and filled them with bright succulents to tie the plantings to the deck.

A public beach access runs alongside the garden, which is just fine with the owners. They placed a bench at the edge of their property so beachgoers can sit and enjoy the view of the ocean and the garden.

FACTS AND FIGURES

ELEVATION 24 feet
ANNUAL AVERAGE RAINFALL 19.25 inches, falling from November through March; dry in summer
ANNUAL RAINY DAYS 32
SUMMER HIGH 85°F, typically in August
WINTER LOW 37°F in December
HUMIDITY Ranges from about 95% in June, July, and September to 63% in December
IRRIGATION Drip
SOIL TYPE Clay
USDA ZONE 10a
DOMINANT PLANT TYPE Succulents

KEY PLANTS FOR COLOR

Aeonium 'Sunburst' and 'Zwartkop'
Agave vilmoriniana 'Stained Glass'
Crassula capitella 'Campfire'
Crassula ovata 'Hummel's Sunset'
Echeveria agavoides 'Lipstick'
Echeveria elegans
Echeveria 'Etna'
Euphorbia characias 'Tasmanian Tiger'
Kalanchoe luciae
Senecio mandraliscae
Yucca 'Bright Star'

Garden Gallery

Colors of the Desert

Tall planes of vertical color surround and cut through Alan Richards's Tucson garden, starting with the privacy walls that separate a streetside garden from his private front yard. Richards painted the walls a yellow-green that he describes as "the color creosote leaves turn just before summer monsoon rains begin."

Visitors to Richards's garden first see those walls as the backdrop for a colorful desert garden of yellow-flowered *Caesalpinia gilliesii* (bird of paradise bush), purple-blush *Opuntia* cactus, shrubby creosote, spikey green *Dasylirion acrotrichum* (desert spoon), yellow-flowered *Parkinsonia florida* (blue palo verde), spiny

In this Tucson garden, tall planes of vertical color divide the streetside garden from the private front yard.

Richards's "desert jungle" has complex layers of foliage in ghostly silver, blue green, pale green, and even apple green.

green *Acacia greggii* (catclaw acacia), and soft green–leaved Australian *Eucalyptus microtheca*. This more typical Tucson plant palette gives no hint to the paradise of color and diversity that lies just beyond the entry gate.

Inside, even more intensely colored walls create a dramatic backdrop for the deep purple *Sophora* flowers that open in late May and the golden yellow of desert wildflowers. Bright turquoise walls add an unexpected punch and contrast. Against those intense colors, Richards planted a "desert jungle" with complex layers of foliage in ghostly silver, blue green, pale green, and even apple green.

This space is Richards's personal Shangri-La, and one of the most colorful and diverse gardens in all of Tucson, thanks to his obsessive love of plants, united with his sense of space and color.

When Richards bought this half-acre property in 2005, neither the small, 1940s brick home, nor its landscape were notable. He started rebuilding with the goal of making the home and garden work together. He describes this new incarnation as "sort of a contemporary fusion…. it's a little streamline, some traditional."

Inside, for example, he removed unnecessary walls to create one large central living area. One wall is almost entirely glass, including the doors that look out into a broad patio surrounded by garden.

Outside, Richards's aversion to straight lines results in many small, interconnected garden spaces. They aren't garden rooms as much as they are garden "experiences." Vivid colored, poured concrete forms bump in and out of winding, wiggly paths that connect one secluded garden space to the next. In each, the blocky shapes form tables, benches, and lounges painted with bright yellows, rich greens, teal blues, and deep indigo purple.

These concrete elements anchor beds packed with Richards's extensive plant collection. Richards is a self-described plantaholic—if a new plant is introduced, he has to have it. Unlike the typical Tucson garden, there is little, if any, bare space between plants here. This overplanting is deliberate. When he finds something new to try, he looks for a place to tuck it in. This approach, he cautions, works only if one is prepared to deal with the pain of selectively editing out plants on a regular basis, especially those plants that grow twice the expected size.

Richards traces his plant obsession to his early childhood, not far from Tucson, where his grandmother first gave him a purple dahlia tuber. He honed his plant addiction during decades of living and gardening in San Diego. Now, back in Arizona, Richards enthusiastically experiments with natives and desert exotics, many new to Arizona horticulture.

Surprisingly, Richards says he draws inspiration from Japanese garden style. "In a Japanese garden you see a scene from

Richards draws inspiration from Japanese garden style, where every viewpoint offers a vignette of layered plants sited for their architectural shapes and textures.

the main house," he explains, and everywhere else you look there is a view, a viewpoint, a vignette. Also like in Japanese gardens, Richards layers plants based on their architectural shapes and textures, but in his desert garden, the plants happen to be agaves with large teeth, golden barrel cacti with big spines, and *Hesperaloe parviflora* (red yucca) with bold shapes.

Garden Gallery

Natives and plants from other desert regions form the backbone of the gardens. Inside the front walled garden, Richards relies on creosote and purple *Opuntia* cacti, along with agaves and yuccas as "thread plants" that repeat throughout the garden. These plants are mostly from the Sonoran and Chihuahuan Deserts and reflect the local horticulture. "Tucson lives with the desert rather than fights with it," Richards explains. "Even in the inner city, native plants are left intact; we have a big horticulture mindset."

Within that backbone, this garden is home to 11 species of dry-growing palms, 14 species of ancient cycads, and many kinds of *Acacia* trees native to deserts of Australia, Africa, and the United States. Each species of *Acacia* contributes a unique color, form, and structure. The upright, weeping *A. pendula* has soft gray leaves that contrast with its mahogany brown trunk; the tall *A. stenophylla* (shoestring acacia) has a high, open canopy of long, narrow, deep green leaves; *A. greggii* (catclaw acacia) has a thorny trunk, fine leaves, and creamy yellow, rod-shaped flowers; cold-hardy *A. schaffneri* (twisted acacia), from New Mexico's Chihuahua Desert, is thorny with almost-purple stems and fine, green leaves.

The rear garden is softer, with an emphasis on finer-textured plants, and even some perennials. Here, flowers play a larger role, with many chosen to attract hummingbirds. There are many different salvias, including several *Salvia greggii* (autumn sage) hybrids with purple, blue, and red flowers. Based on his years of experience gardening near the coast, Richards also experiments with *S. clevelandii* (Cleveland sage) and its hybrids. These

Richards repeats plantings of the tiny green-leaved *Larrea tridentata* (creosote bush), purple-paddle *Opuntia* cacti, as well as long, narrow-bladed agaves and yuccas to tie the garden together.

aromatic-leaved San Diego County natives bloom lavender to deep purple in spring.

Flowering shrubs and bulbs add to the nearly year-round bloom color that Richards describes as being present in "doses, spots, and splatters." He likes color contrasts, especially bright yellow next to silver. Toward that end, there are mounds of diminutive yellow-blooming *Chrysactinia mexicana* (Damianita daisy), slightly larger yellow *Berlandiera lyrata* (chocolate

Pathways weave through the rear garden, where there's a softer look created by finer-textured plants and flowering perennials, along with art pieces and found objects.

flower), and golden yellow *Baileya multiradiata* (desert marigold). Nearby, a large, blue-leaved agave forms the backdrop for the purple Midwest native *Glandularia bipinnatifida* (prairie verbena) and the sherbet orange and watermelon pink flowers of globe mallow (*Sphaeralcea*). The combinations are beautiful.

Richards's most special plants go into pots on the patio that serves as a broad, wide plane outside the home's glass wall. Pots too are chosen for their color: browns, black, greens, and blues, and every once in a while, according to Richards, a yellow pot "slips in."

Choosing new plants is simple since the garden's color palette is so well established. "If I am out shopping and I see a yellow flower I like," Richards says, "I nab it."

Irrigation of course is an issue in this garden. Richards graded his property with berms and channels to direct rainfall and keep water on site. In the streetside public landscape, trees were planted first and irrigated with bubblers for two years. As the smaller plants went in, Richards made

Garden Gallery

A pot converted to a water feature in an intimate corner of the rear garden attracts birds and makes a lovely burbling sound.

Upright spires of blue-flowering *Salvia chamaedryoides* (germander sage) echo the shapes of nearby columnar cacti.

wells around each one to catch rainwater. He also added drip irrigation. At year five, Richards started to reverse the process with the goal of weaning trees off supplemental irrigation. He is still adding new lines so he can better control how much irrigation goes to each area of the garden. His goal is to reduce water use and water costs. "Every year for the last four years," he says, "I've reduced the water I use by 10%. Now, I'm at 40% reduction and I'm about to add new lines so I can water smarter." Tucson water is so expensive that Richards doesn't mind spending ten dollars on a new irrigation valve since it allows him to divide the garden into more areas, then water each on its own schedule.

This kind of flexible thinking is typical of Alan Richards's approach. Both his home and garden change regularly. "You can't live with a hard plan when you are a plant collector," he explains. "You find something new and say, 'well it will fit in here,' and then a different plant doesn't seem so special

The rear garden features red-flowering *Salvia greggii* with purple-flowering *Glandularia bipinnatifida* and golden yellow-blooming *Chrysactinia mexicana*.

Yellow, daisylike *Berlandiera lyrata* is one of the stalwart's of the desert garden.

Garden Gallery

A large, blue-leaved agave forms the backdrop for the orange and pink flowers of globe mallow.

FACTS AND FIGURES

ELEVATION 2,383 feet
ANNUAL AVERAGE RAINFALL 11.6 inches; peaks in August summer monsoons, with winter rain December through March
ANNUAL RAINY DAYS 27
SUMMER HIGH 102°F, typically in July
WINTER LOW 38°F in December
HUMIDITY Peaks at around 90% in February, June, and August; dips to 70–75% in January
IRRIGATION Drip and hand watering
SOIL TYPE Clay loam
USDA ZONE 9a
DOMINANT PLANT TYPE Mixed succulent and broad-leaved, mostly desert natives

KEY PLANTS FOR COLOR

Achillea 'Moonshine'
Buddleia marrubifolia
Parkinsonia 'Desert Museum'
Euphorbia rigida
Lagerstroemia species
Peniocereus greggii
Ruellia peninsularis
Sophora secundiflora 'Silver Peso'
Trichocereus hybrids

anymore so I get rid of it and replace it with something else." Not even the wall colors are sacred: "If I get tired of the color," he says, "it's twenty dollars of paint and an hour of work to change it…. There are no two walls inside or two walls outside that are the same color, all are meant to be changed at whim."

As Richards talks, he glances down at his shirt and realizes it is the same blue as the wall he is standing next to. He smiles. "These are the colors I look good in! It's a harmony thing. I tell people you need a harmony of color to make things comfortable and flow. You need the same thing with plants, so I have a spikey plant and a round soft plant and put those next to each other to contrast and punch, then put something bright green in front of the gray and it works!"

Garden Gallery

The Art of Borrowed Color

The view from Bob and Judy Schumann's Tucson-area living room is stunning. An enormous picture window frames a view of massive mountains in the background and an equally impressive garden in the foreground. The composition epitomizes the Asian principle of the borrowed view, where the garden is designed to appear as a united whole with the natural landscape.

Here, as in traditional Asian landscapes, the mountains are the focal point. During the day, the Santa Catalina Mountains look chocolate brown against a wide, cornflower blue–colored sky that reflects in the garden's aqua blue swimming pool. At sunset, the mountains turn raspberry red, while the sky and pool glow gold and yellow. Taupe-colored stucco walls surround the garden, interrupted at interesting angles by large panes of steel, all rusted to tones of deep copper. Chocolate, blue, raspberry, gold, taupe, copper—as long as

The view from Bob and Judy Schumann's Tucson-area living room takes in both the surrounding mountains and their Scott Calhoun–designed garden.

there is light, layers of color set the stage for this desert garden.

Nature created the mountains, but Tucson garden designer Scott Calhoun created the garden. Calhoun's approach to the 2,800-square-foot space is a carefully chosen combination of plants, hardscape, and decorative elements, some of which were made or selected by the homeowners themselves. The garden they created honors the borrowed view and Arizona's ever-present sunlight.

The swimming pool, walls, flagstone decks and pathways, as well as a dry streambed, were already in place when Calhoun arrived on the scene. His main task was to assemble and choreograph a carefully chosen plant palette of native and nonnative plants, all adapted to a low-water diet. And, of course, the owners wanted color.

Calhoun learned early in his career not to count on flowers for garden color. "In the more naturalistic design work I do," he explains, "I'm not worried about flower color since flowers riot briefly then lie low." With flower color so ephemeral, Calhoun relies on foliage for the texture and color palette. He selects trees and evergreen shrubs that have the small leaves or fine textures typical of drought-tolerant plants. That kind of foliage can get lost, Calhoun says, "so my tendency is to use strong colors, even earth tones, to bounce the plants off of visually." In the Schumanns' garden, the deep-toned garden walls and their rusted steel panels play that role perfectly.

Calhoun's favors silver foliage, which is a common adaptation for plants that grow where sunlight is intense. The silver, from fine hairs or from a waxy coating on the leaf surface, reflects sunlight and helps

The dry streambed was already in place when Calhoun designed the garden. He embellished the stones with plants that are tough, colorful, and beautiful.

OPPOSITE TOP: At sunset, the garden's pool turns dark, the sky glows gold and yellow, and the mountains' raspberry tones reflect in the home's windows.

OPPOSITE BOTTOM: Stucco walls surround the garden, interrupted at interesting angles by large panes of steel, all rusted to tones of deep copper.

Fouquieria splendens offers "seasonal flower explosions" of color when it erupts in masses of fiery red flowers.

slow water loss from the leaves. Calhoun adds plants with olive green, deep green, blue-green, and even pale purple foliage for contrast.

Against the dark tones and under intense sunlight, Calhoun finds that the smallest spot of bright flower color stands out. *Sophora secundiflora* 'Silver Peso' (Texas mountain laurel) is one example. The designer set these small evergreen trees against the backdrop of deep-toned metal panels. All year, their silvery white-flocked leaves make for a dramatic contrast against the chocolaty rusted metal. When the trees bloom in early spring, they erupt in stunning sprays of intensely violet flowers that smell like grape bubblegum.

Fouquieria splendens (ocotillo) is another native plant with what Calhoun refers to as "seasonal flower explosions." Its upright stems stand 10–15 feet tall, each clothed in clusters of small, olive green leaves that alternate with wickedly sharp spines. In spring, the branch tops erupt in pointed masses of fiery red flowers. The plant's common name, ocotillo, is fittingly derived from the native Aztecan word for "torch."

Ferocactus pilosus (Mexican lime cactus) features a perennial translucent halo of blood red spines, while *Echinocactus grusonii* (golden barrel cactus) has its own halo of translucent yellow spines. When the sun hits at just the right angle, the spines look luminous.

In other dry climates, it is easy to assemble a plant palette where the garden blooms ten or twelve months out of the year. It is much harder in the desert. These gardens have highs and lows as plants go in and out of dormancy in rhythm with their natural cycles. So, for example, the end of April is usually when the desert perennials

Who says that cacti are not colorful? Blazing orange, yellow, red, and pink flowers are frequently on display, presumably a strategy for attracting pollinators in the vast desert.

Calhoun designed square metal planters to sit atop cylindrical gabion pedestals filled with round cobble. He chose rounded cacti for the planters, including small, spiny *Thelocactus*.

bloom, Calhoun says. Next come the cacti that flower in blazing orange, yellow, red, and pink, a strategy presumably for attracting pollinators in the vast desert.

"May and June bring blooms like Arizona queen of the night (*Peniocereus greggii*). Then the saguaros (*Carnegiea gigantea*) bloom and fruit," Calhoun says. "By June, most flowering is done. Plants hunker down and wait for rain, or dry up and die, waiting for the next fall." Calhoun works with these cycles rather than fight against them. For each garden, he specifies custom mixes of wildflower seeds, designed to bring two waves of seasonal color, each lasting about two weeks. The first wave comes in March and April, at the tail end of winter rains; the second wave arrives in August and September following summer monsoon rains. His mixes include annuals and perennials that bloom in colors like cobalt blue (*Phacelia campanularia*, desert bluebell), golden orange (*Eschscholzia californica* subsp. *mexicana*, Mexican gold poppy), and coral red (*Penstemon superbus*, superb penstemon). He encourages clients to let the flowers go to seed. Season after season, these wildflowers deposit a robust seed bank that ensures waves of color in successive years.

Beyond the plants, Calhoun designed a series of contemporary decorative planters to complement the Schumanns' contemporary-style home. The homeowners are avid do-it-yourselfers with expertise in woodworking and metalworking, so they took Calhoun's planter design and went to work. They used metal reinforcing wire to make the gabion pedestal cages, and then fabricated the 2-foot square planters from 12-gauge hot rolled steel. The square metal planters each sit atop the cages,

Garden Gallery

Bob Schumann's wood and metal bee habitats bring some much-appreciated verticality to the garden.

which are between 20 inches and 2 feet tall, all filled with round cobble.

After the planters were installed, Calhoun planted rounded cacti into each one; purple-bladed *Opuntia violacea* var. *macrocentra* (tuxedo spine prickly pear), *Astrophytum myriostigma* (bishop cap), small clusters of *Thelocactus*, and so on. At the base of each gabion he created contrasting color and texture by planting blue-tone, grasslike *Calibanus hookerii* (gorilla's armpit), pink-blooming *Penstemon triflorus* (hill country penstemon), and *Euphorbia antisyphilitica* (candelilla), an upright, pencil-thin succulent with tiny pink flowers. The rusted metal planters in the center of the garden are a visual connection to the rusted panels in the surrounding walls.

The homeowners contributed in other ways as well. They fabricated metal and wood end tables, a beautiful large outdoor dining table, and metal wall sculptures. Bob Schumann made a rusted metal outdoor shower enclosure that matches the wall panels and gabions. He also made a trio of freestanding wood and metal bee habitats that double as garden sculptures— all fabricated in the couple's garage.

Judy Schumann selected accessories to enhance the garden's color scheme. The small patio outside the master bedroom features pillows and a wall hanging whose golden yellow tones match the color of the fruit of the nearby *Ferocactus wislizeni* (fishhook barrel cactus). She accessorized the main patio in terra cotta and oranges that echo the rusted metal. Dusky purple in pillows and ceramic orbs picks up the purple tones in the taupe-colored walls, along with the blue-purple blades of *Opuntia violacea* var. *macrocentra* and apricot-blooming *Stapelia grandiflora* (African starfish flower), a

Garden Gallery

This inviting patio corner pulls in the colors of sunset—red, orange, and dusky purple—in plants, furniture, and accessories.

succulent whose tentaclelike branches spill from a ceramic pot.

The process of creating this garden took the Schumanns, Kansas City transplants, on a journey to a different mindset, under the direction and tutelage of Scott Calhoun. The couple adores their new garden for its beauty and also for how easy it is to care for. The garden rarely needs more than the occasional pruning and deadheading, marvels Judy Schumann. "I think of what we did in the Midwest," she says, "all that mowing and fertilizing…. how nice not to have to do that stuff in the spring anymore!"

The small patio outside the master bedroom features pillows and a yellow-toned wall hanging that matches the fruit of the nearby *Ferocactus wislizeni*.

FACTS AND FIGURES

ELEVATION 2,842 feet
ANNUAL AVERAGE RAINFALL 11.5 inches; summer monsoons peak in August with winter rain from October through March
ANNUAL RAINY DAYS 26
SUMMER HIGH 103°F, typically in July
WINTER LOW 38°F in January
HUMIDITY Peaks around 90% in February, June, and September; dips to 70–75% in January and April
IRRIGATION Drip
SOIL TYPE Granitic
USDA ZONE 9a
DOMINANT PLANT TYPE Desert succulents and broad-leaved plants

KEY PLANTS FOR COLOR

Calliandra 'Sierra Starr'
Dalea capitata
Eschscholzia californica subsp. *mexicana*
Ferocactus pilosus
Ferocactus wislizeni
Fouquieria splendens
Mascagnia macroptera
Opuntia violacea var. *macrocentra*
Pedilanthus macrocarpus
Penstemon triflorus
Sophora secundiflora 'Silver Peso'

Garden Gallery

Hot Color, My Garden

Directions to my home end like this: "At the end of the street, look to your left. Our house is terra cotta colored with a green metal roof and a bright-colored garden of plants you've probably never seen before. Oh, and there's no lawn."

When we bought this property in 1992, it was a long-neglected home on a two-thirds–acre corner lot, which was unfenced and largely unlandscaped. The backyard had a flowering mulberry tree, a few dying citrus trees, a line of spikey yucca, and a mini grove of stunted Christmas trees. The only plants of value were three enormous black acacia trees that dominated the front yard, an impressively large Torrey pine in back, beneath which grew a pair of fruiting pomegranates.

Our initial focus was the backyard, since that's where we spent most of our time with our then-small children. Landscape designer Linda Chisari did the inital garden layout but several years passed before we turned our attention to the front garden, and even then, it was a slow process. We cleared out old plants and removed a cracked asphalt driveway. Then, we collected piles of salvaged broken concrete for building retaining walls, stairs, and walkways.

In 1996, a large fire singed the edge of our neighborhood. It fortunately spared

When I tell visitors how to reach my home, I tell them to look for the terra cotta house with a green roof and a garden of plants they've probably never seen before—and no lawn.

our home, but it was a loud wake-up call to replace our highly flammable, wood shingle roof. Since we were reroofing, we decided to restucco our home's cracked and crumbling walls. That was my opportunity to choose a new house color, as well. Rather than keeping with bland white, I chose rich terra cotta to set the stage for my dream garden.

Even then, I was determined to create the most beautiful garden that would use the least amount of water. I had already amassed a sizeable collection of aloes, agaves, and other succulents. My home "nursery" included *Parkinsonia* 'Desert Museum' (palo verde), which I brought back from a trip to the desert. Today, this tree is a mainstay of waterwise gardens, but in 2000 it was largely unknown in Southern California. In fact, mine is probably the first one planted in San Diego County.

Along with the succulents and desert natives, I was enamored with flowering shrubs in the family Proteaceae, including leucadendrons and grevilleas. I knew them as staples of San Diego's cut flower industry, where they grow in the thinnest soils, on steep hillsides, on just two or three gallons of water a week. They had yet to become popular landscape plants, but I followed my intuition to plant them with my succulents.

I was also interested in natives. Clearly the plants that evolved here are those that grow best with the limited water that nature provides. The bright blue–flowering *Ceanothus* (California lilac), golden yellow–flowering *Fremontodendron californicum* (California flannel bush), and the huge,

Garden Gallery

Terra cotta walls and forest green trim enhance the springtime explosion of colors and textures in my garden.

white- and yellow-blooming *Romneya coulteri* (matilija poppy), were just a few of the natives on my wish list.

Beyond that, though, I had no idea where to start. Collecting plants, writing about gardens, and designing gardens are different sets of skills. So I called on my friend, plantsman and landscape designer Scott Spencer, for help. Spencer looked at my plants, listened to my goals, and surveyed my space. Then, we went shopping for even more plants.

The garden we created was unlike anything in our community.

Spencer used my desert museum tree as the garden's centerpiece. Each spring,

Desert museum palo verde acts as the centerpiece of the front garden. It blooms on and off through the year, but in spring explodes with bright yellow and orange flowers.

it explodes with bright yellow and orange flowers that glow against the terra cotta background. We underplanted the tree with *Salvia chamaedryoides* (germander sage), which has tiny, round gray leaves and cornflower blue blooms almost every day of the year, whether irrigated or not. Spots of sunny yellow–flowering *Calylophus drummondianus* (sundrop) edge the garden beds.

Intense, saturated colors like these soon filled the garden. We used red-leaved conebushes (*Leucadendron*), burgundy-leaved *Cotinus coggygria* (smoke tree), red-flowering *Calliandra californica* (Baja fairy duster), and bright orange–flowering California native poppies. Then, as now, aloes bloom in winter and spring. My favorite, *Aloe rubroviolacea*, blooms in November and December with tall candelabras of showy dusky coral flowers against blue-green blades that blush violet. In early summer, the octopus-like, smooth green blades of A. camperi are topped in spikes of apricot-orange blooms.

The garden is red and yellow, blue and green, burgundy and chartreuse. I resisted pink until Spencer insisted on planting an African daisy, Arctotis 'Big Magenta', that forms a broad, low mat of silvery leaves topped in deep magenta flowers, each 3–4 inches across. A few years later, I added Arctotis The Ravers 'Pink Sugar', whose deep pink flowers are pale orange in the center. It is an unusual, eye-catching color combination, especially in contrast to the silver-gray foliage.

Like all gardens, this one changes over time. I constantly add new plants as I encounter them and remove plants that fail or simply don't work in the space. I have an ongoing experiment with different combinations of leaf shape, leaf color, and flower color.

One of my favorite plants is Bougainvillea 'Bengal Orange'. Even though I find bougainvilleas to be way overused in Southern California gardens, I love the cream and green variegated foliage of 'Bengal Orange'. This is a smaller cultivar that reaches only 4 feet tall and 8–10 feet across. Its blooms (technically bracts) emerge salmon colored, then age to dusky pink. I planted it under the 'Desert Museum' tree and added Alstroemeria 'Third Harmonic', a Peruvian lily whose flowers are orange with burgundy and yellow markings in the throat. The orange matches an orange shade in the Bougainvillea flowers. In combination with bright yellow–flowering Hunnemannia fumariifolia (Mexican tulip poppy), Calylophus

Foliage brings the colors of cabernet and merlot to the garden: Cotinus coggygria, Leucadendron 'Ebony', and this Cordyline 'Red Sensation'. In spring, the pokerlike flower clusters of Melianthus major (honeybush) bloom the same color.

ABOVE: Among several aloes in my garden, *Aloe rubroviolacea* is my favorite; it blooms starting in early winter with tall candelabras of coral flowers against blue-green, violet-blushed blades.

LEFT: *Aloe camperi* is topped with apricot-orange blooms in early summer.

South African bulbs are important components of the early spring garden. Here, a dwarf pink bugle lily (*Watsonia*) blooms in coordination with orange-flowered *Glaucium flavum* (horned poppy) and *Arctotis* 'Big Magenta'. Green, gray-green, and blue-green leaves, ruffled, narrow, and wide, enrich the tapestry.

drummondianus and succulent *Senecio mandraliscae* (blue chalk fingers), the garden's central focal point is a riot of color from late winter well into summer.

One corner of the garden features *Sphaeralcea ambigua* 'Louis Hamilton' (desert mallow), which has fuzzy silver-green leaves and watermelon red flowers. A friend gave me a tall, unusual yellow-flowering montbretia (*Crocosmia*) that blooms at the same time. The thick, succulent blades of *Aloe ferox* and three very round golden barrel cacti (*Echinocactus grusonii*) add to the textures. At the same time, the veil of translucent yellow spines on the cacti and the blue-gray tones of the aloe enhance the color scheme.

I take great delight in playing with rich, intense colors that are beautiful on their own, yet so much more impressive in combination with others. I layer the colors to create "color intersections." Yellow is pretty and so is pink; put them together, though, and each is that much more vibrant. Add sky blue and deep green and you have a stunning combination.

Against white walls, however, even the brightest colors would look lackluster. As I'd hoped, the terra cotta background adds dimension and contrast to the colors of the plants. I could have created the same effect using chocolate or latte or deep sage green or another intense earth tone for the walls.

Time is another dimension of garden color. Colors come and go through the season. The blue-green leaves of *Eschscholzia californica* (California poppy) emerge in late winter and are topped in bright orange flowers through spring. I never quite know where they will sprout, but they are welcome everywhere. Later in spring, the round leaves of *Cotinus*

The garden's central focal point is a riot of color from late winter well into summer, thanks to the combination of dwarf *Bougainvillea* 'Bengal Orange', *Senecio mandraliscae*, and yellow-flowered *Calylophus drummondianus*.

Flower color echoes: *Alstroemeria* 'Third Harmonic' with *Bougainvillea* 'Bengal Orange'.

Garden Gallery

In one corner of the garden, red-flowering *Sphaeralcea ambigua* 'Louis Hamilton' sits next to a tall, yellow-flowering *Crocosmia* and the thick blades of *Aloe ferox*.

coggygria emerge deep burgundy, then turn blue-green toward the end of summer. In fall, they turn crimson and apricot before they drop. Nearby, the leaves of *Leucadendron* 'Ebony' are a similar burgundy but so dark that they look almost black. While *Cotinus coggygria* is deciduous, *Leucadendron* 'Ebony' is evergreen; so burgundy is represented year-round but more prominent from spring through fall. As it happened, *Cotinus coggygria* was planted in a spot where the leaves are backlit in fall. With the sun low in the sky, the beautiful colors glow in the late afternoon sun.

Spring and fall are definitely the most colorful times in my garden. Many of the plants bloom in the shorter days when the air and soil are warm but the air is still cool. When summer comes and the garden takes its siesta, I look to the vegetable garden for color. Red tomatoes, yellow peppers, purple eggplants, golden squash, and green zucchinis are so beautiful that I hesitate to pick them.

While the vegetable garden does need more water than the ornamental plants, it takes up only a fraction of the overall garden when it comes to space. And, the water I "spend" on vegetables feeds my family, as do the fruits that grow on trees irrigated with gray water from our washing machine.

Since the beginning, my garden has been watered with in-line drip irrigation, which is both efficient and reliable. I also find it to be the easiest to maintain for both ornamental and edible plants. No heads blow off or get kicked over, no lines burst apart; no emitters get lost or moved—it just works. When the rains begin in November, I turn the irrigation system off, save for the occasional deep soak during hot, dry Santa

Color and texture from arid- and Mediterranean-climate plants: *Agave attenuata* and narrow-stemmed *Pedilanthus bracteatus* from Mexico, tall *Cordyline* 'Red Sensation' from Australia, North American native *Muhlenbergia capillaris*, South African coral-colored bugle lilies (*Watsonia*), and succulent *Aeonium* rosettes from the Canary Islands, all growing beneath a large *Schinus molle* pepper tree from the Andes of Peru.

Ana winds or long dry periods in winter. It usually stays off until March or April.

That said, some garden beds are no longer irrigated at all. Native plants, those from the deserts, and many of the Mediterranean-climate plants I watered for the first few years are now watered only by rainfall. I'm sure the plants are helped by the extra water that our rain gutters direct into the soil rather than into the street, and by the leach field from our home's septic system. Talk about recycling!

Since our property is close to a creek, the soil is very sandy. Sandy soils drain well but lack organic matter so they have a hard time holding water near where plant roots need it. After decades of my layering mulch onto the soil, our soil now both drains and holds water in perfect balance.

It takes time to create a garden, especially one done on a shoestring and with lofty goals. When I stand back and look at what we've created, the garden has more than achieved my goals. It serves as a laboratory to test ideas both large and small; I use it in my writing, my teaching, and my work designing gardens for others.

What makes me most proud, though, is the appreciation from my daughter, who once announced, "Mom, we have the coolest garden in the neighborhood!"

FACTS AND FIGURES

ELEVATION 16 feet
ANNUAL AVERAGE RAINFALL 12 inches, mostly October through March; no summer rainfall
ANNUAL RAINY DAYS 24
SUMMER HIGH 94°F, typically in August
WINTER LOW 33°F in December, with winter nights dipping down to 26–27°F
HUMIDITY Peaks in August and November at more than 90%, dips to 70–75% in January, March, and October
IRRIGATION In-line drip
SOIL TYPE Sand and sandy loam
USDA ZONE 10a
DOMINANT PLANT TYPE Mixed Mediterranean with succulents, California, and desert natives

KEY PLANTS FOR COLOR

Aloe cameronii
Aloe rubroviolacea
Alstroemeria 'Third Harmonic'
Anigozanthos 'Big Red'
Arctotis The Ravers 'Pink Sugar'
Bougainvillea 'Bengal Orange'
Calylophus drummondianus
Cotinus coggygria
Eschscholzia californica
Hunnemannia fumariifolia
Leucadendron 'Ebony'
Leucadendron salignum 'Winter Red'
Leucospermum 'Veldfire'
Leucospermum cordifolium 'Yellow Bird'
Melianthus major and *M. major* 'Purple Haze'
Salvia chamaedryoides
Sphaeralcea ambigua

PLANT DIRECTORY

The colorful gardens in these pages grow across a vast region, from Albuquerque in the east, to San Diego in the west, and north to Santa Rosa. Though all of these gardens experience long, rainless periods during the year, gardening conditions vary significantly. Plants featured in this section are drawn from these showcase gardens. Some grow well throughout the entire region, while others thrive in a far narrower range. All, however, are worth trying.

You'll find that this list is skewed toward Mediterranean-climate plants for a simple reason: the diversity of Mediterranean-climate plants is astounding. Only 2 percent of the earth has a Mediterranean climate, yet 20 percent of all plants are endemic to those regions. Those wild plants are the sources for our garden plants. When you add to the Mediterranean-climate plants the many desert- and temperate-climate natives that tolerate dry conditions, you soon realize the enormity of the plant palette for waterwise gardens.

If this list omits some of your region's tried-and-true drought-tolerant plants, be assured that's intentional. My goal is to expand your plant horizons and steer you toward some lesser-known plants. These are plants that are still readily available, though you might have to ask your local independent nursery to order them for you.

I encourage you to experiment with new plants and embrace failure. In my garden, I have a "three strikes, you're out" rule. When I find something new, I always buy three plants, in part to increase the possibility of at least one surviving. I plant them all together in my garden. If the first plantings fail, I try again, this time placing the plants in a spot where the growing conditions are a little different. If that planting fails, I'll try a third time, again in a different location. If the third planting fails, I know that plant doesn't belong in my garden and I move on to something else.

That said, if it is a plant I *truly* covet, after the third strike I'll purchase just one plant and grow it in a pot. That way I can experiment with growing conditions, with soil mixes, with the amount I water it, and so on. Sometimes, I discover why the plants failed in the ground and give it another go. Other times I end up with a lovely potted plant, which is just fine with me.

Understanding Your Climate

In California, gardens are referred to as Mediterranean-climate gardens, based on the rainfall pattern. Like in the Mediterranean, most of California's precipitation occurs from November to March or early April. The hottest months of the year are very, very dry. It's a great climate for people, but it can be difficult for plants unless they've evolved here or in one of the other four Mediterranean-climate regions of the world.

In Arizona, rainfall (and snowfall in some regions) comes in both winter and summer. In New Mexico, rains are concentrated in summer. The total precipitation (rain plus snow) in Phoenix, Tucson, or Albuquerque is nearly the same as the rainfall in Los Angeles or San Diego. But the fact that the rains (and snow) happen in different seasons means that some desert plants in Mediterranean-climate gardens require summer irrigation.

On the flip side, California as a whole is much more humid than New Mexico or Arizona (save for the California desert, of course). Plants that thrive in California shrivel in more arid climates, or freeze in the high-desert winters.

So while all of these regions can be described as "dry" or "arid," their plant palettes are not identical. Many plants thrive in the entire Southwest. Others are more regional.

Achillea

yarrow

Yarrows are great for a soft-looking, low evergreen edging or meadow plant. They have green or gray ferny foliage that grows as low mounds, rosettes, or mats. They produce many tall flower stems topped with umbrella-shaped flower clusters comprised of dozens of tiny flowers. There are different species and cultivars to choose from, some native to the West, others from Europe and Asia. In general, yarrows grow in full sun or light shade, well-draining soil, and require only infrequent watering once established. Butterflies frequent yarrow.

Achillea 'Hella Glashoff'

Hella Glashoff yarrow

COLD TOLERANCE 0°F or colder
INTEREST spring to summer
SIZE 18 in. mounds
COLOR green, soft yellow
PERENNIAL

This *Achillea filipendula* hybrid has green foliage and soft yellow flowers.

Achillea 'Hella Glashoff'

Achillea millefolium var. californica
western yarrow

COLD TOLERANCE 0°F or colder
INTEREST spring to summer
SIZE 1–2 ft. tall, 2–3 ft. wide
COLOR green with white
PERENNIAL

Western yarrow is a low-growing California native with mounds of ferny green leaves. Grow in full sun along coast, or some shade inland.

Achillea millefolium var. *californica*

Achillea millefolium 'Paprika'
paprika yarrow

COLD TOLERANCE 0°F or colder
INTEREST spring to summer
SIZE 2 ft. × 2 ft. mounds
COLOR green, yellow, red, salmon, cream, brown
PERENNIAL

Unusual red flowers make this yarrow distinct from the others. Each has a yellow center and the ferny foliage is green. As the flowers age, they turn salmon, cream, then brown.

Achillea millefolium var. rosea 'Island Pink'
Island pink yarrow

COLD TOLERANCE 0°F or colder
INTEREST spring to summer
SIZE 1–2 ft. tall, 2 ft. wide
COLOR green with rose pink
PERENNIAL

This native was collected from the islands off California's coast. Grow in full sun with some shade.

Achillea millefolium 'Paprika'

Achillea millefolium var. *rosea* 'Island Pink'

Achillea 'Moonshine'

moonshine yarrow

COLD TOLERANCE 0°F or colder
INTEREST spring to summer
SIZE 2 ft. × 2 ft. clumps
COLOR silver, bright yellow flowers
PERENNIAL

Achillea 'Moonshine' is a hybrid whose ferny foliage is broader than most other garden yarrows and silver rather than green. Grow in full sun.

Aeonium

Aeoniums are known for succulent rosettes that look like ever-present flowers. Some grow atop stout succulent stems, others atop narrow stems. Some are single stems topped in a single rosette, some clumping; others are more like small shrublets with many stems and rosettes. Colors range from green to almost black, some reddish, some variegated. Heights range from just 1–4 feet, depending on variety. Aeoniums are grown for foliage, not for the long stalks of tiny yellow flowers (Aeonium nobile is an exception). The flowering rosettes die after flowers fade but other rosettes remain. These are generally winter growers that prefer little if any summer water. Aeoniums are extremely versatile in the garden, in containers or in the ground, as edging or specimens, and used for form, shape, color, and interest. They do best in full sun to light shade, depending on variety.

Achillea 'Moonshine'

Aeonium 'Cyclops'

COLD TOLERANCE 25°F
INTEREST year-round
SIZE 3–4 ft. tall, 1 ft. wide
COLOR bronze-red and green
SUCCULENT

Succulent rosette with shiny red-bronze blades. New blades often emerge green, so rosettes appear to have green centers. Rosettes sit atop stems, though the occasional rosette will sprout on the side of the stalk. These aeoniums do best in full coastal sun to light shade. If rosettes turn green, they are planted in too much shade. Grow in well-drained soil, and irrigate occasionally, very little in summer. Flowers are yellow.

Aeonium 'Kiwi'

COLD TOLERANCE 25–30°F
INTEREST year-round
SIZE 2 ft. mounds, shrublet
COLOR yellow, green, pink
SUCCULENT

Small shrublet succulent with variegated rosettes of tricolored leaves: yellow toward the center, greener toward the outer, all edged in deep red. Prefers cool sun or light shade in Mediterranean climates.

Aeonium 'Cyclops'

Aeonium 'Kiwi'

Aeonium nobile
noble aeonium

COLD TOLERANCE 20–25°F
INTEREST summer and year-round
SIZE 1–2 ft. tall, 1 ft. wide
COLOR pale olive green, coral, vermillion, yellow
SUCCULENT

This is one of the only common garden aeoniums with a notable flower. In summer, the rosettes feature stout flower stalks, each topped in a mass of tiny, vermillion, star-shaped flowers with yellow anthers. The leafy rosettes are unusual too. They are large with fuzzy, triangular blades that are pale olive green with reddish edges and very fleshy. When heat and water stressed, blades turn coral, almost orange. One rosette atop a single stem.

Aeonium 'Sunburst'

COLD TOLERANCE 30°F
INTEREST year-round
SIZE 2–3 ft. tall, 1 ft. wide
COLOR yellow, green, pink
SUCCULENT

These succulent rosettes usually form as a single head atop each stem. Rosettes are made up of green and pale yellow–striped blades edged in pink. Together the blades make the rosette look like a colorful pinwheel. Grow in full sun along the coast, or part sun in hotter climates.

Aeonium nobile

Aeonium 'Sunburst'

Aeonium 'Zwartkop'

COLD TOLERANCE 28°F
INTEREST year-round
SIZE 2–3 ft. tall, 1 ft. wide
COLOR burgandy-black
SUCCULENT

The succulent rosettes of this aeonium form as single heads on succulent stems. Each rosette is made up of shiny blades that range from claret to deep burgundy, almost black. They provide an excellent color accent in the garden. Grow in full sun along the coast, part sun in hot climates. If rosettes turn green, the plants are in too much shade.

Agave

Agaves are succulents from the Western Hemisphere, all in variations of vase to rosette shapes, and growing 1–12 feet tall and wide, depending on the species. They are grown for their foliage, which can range from bright green, ice green, blue-green, or silver-green, to variegated with cream or yellow or sometimes edged in red or copper. Many have toothed edges and terminal spines; a few are smooth and spineless. Most make pups (sideshoots). Flowering occurs after many years and the main plant usually dies afterward. Flower stalks are typically very tall and either straight or branched, with pale cream to yellow flowers. Flowers are filled with nectar that attracts bees and hummingbirds. Some species' flowers turn to bulbils on the stalk, tiny plantlets that root once the stalk falls to the ground. Most agaves are extremely drought tolerant, adaptable to a range of soils, and generally tough plants.

Aeonium 'Zwartkop'

Agave americana var. mediopicta 'Alba'

white-striped century plant

COLD TOLERANCE 15–20°F
INTEREST year-round
SIZE 3–4 ft. tall, 4–6 ft. wide
COLOR blue-green and pale ice green
SUCCULENT

This is a Mexican native whose leathery blue-green blades have pale, ice green stripes down the centers. Give plants plenty of space. Edges are very spiny so site far from walkways or high-traffic areas. Mother plants will die after they bloom, but this happens only occasionally. Like other agaves, this one pups. Grow in full sun or part shade.

Agave applanata 'Cream Spike'

COLD TOLERANCE 15°F
INTEREST year-round
SIZE 6 in. tall, 4 in. wide
COLOR green, cream, and brown
SUCCULENT

'Cream Spike' is a mini agave, but when massed, forms a colony of variegated plants, each with succulent blades that are green down the center with creamy, yellow-white edges and a long, dark brown spike at the tip. It does well in sun or shade. Native to Mexico.

Agave americana var. *mediopicta* 'Alba'

Agave applanata 'Cream Spike'

Agave attenuata 'Nova'
blue foxtail agave

COLD TOLERANCE 25–30°F
INTEREST Year-round
SIZE 3–4 ft. tall and wide
COLOR blue-green or blue-gray
SUCCULENT

Blue foxtail agaves form colonies of smooth-leaved rosettes. The straight species has icy green blades and a curved flower stalk, while 'Nova' has wider bluish blades and an upright stalk. Blade edges are smooth and without spines or teeth. Grow in full sun or light shade. Sometimes called 'Boutin Blue'.

Agave 'Blue Flame'

COLD TOLERANCE 20–25°F
INTEREST year-round
SIZE 5 ft. tall, 5–6 ft. wide
COLOR blue-green blades
SUCCULENT

The overall shape of this agave is flamelike, hence its name. It forms 5-foot rosettes whose waxy surface and deep blue-green blades feature distinct striations from leaf tip to base. Edges are coppery red and finely serrated, and each curved blade tip has a spine. Grow in full sun along the coast and light shade (especially afternoon shade) in inland and desert climates. Plants multiply by forming pups, which eventually form masses many feet across.

Agave attenuata 'Nova'

Agave 'Blue Flame'

Agave 'Joe Hoak'

COLD TOLERANCE 15°F
INTEREST year-round
SIZE 2 ft. tall, 2–3 ft. wide
COLOR bright green and yellow
SUCCULENT

This beautiful agave has gracefully curved blades that are striated yellow and bright green. The blades are greener in the center and more yellow toward the edges. 'Joe Hoak' looks great in a cobalt blue or oxblood red glazed container, or planted in the ground against a backdrop of deep green. It can light up a slightly shady spot.

Agave 'Kissho Kan'

COLD TOLERANCE 30°F
INTEREST year-round
SIZE 1 ft. tall and wide
COLOR blue-green, cream, and copper
SUCCULENT

Succulent blue-green blades have cream-colored striations and cream white margins. Each succulent blade is edged in copper-colored teeth. This slow-growing beauty is great in a container, especially if the container is glazed in a copper color. It does best in full or part sun and requires little water. Also called *Agave potatorum* 'Variegata'.

Agave 'Joe Hoak'

Agave 'Kissho Kan'

Agave vilmoriniana 'Stained Glass'

variegated octopus agave

COLD TOLERANCE 20–25°F
INTEREST year-round
SIZE 4 ft. tall and wide
COLOR bright green and yellow
SUCCULENT

Octopus agave is a vase-shaped Mexican native with gracefully arched, octopuslike blades that are deep green with pale yellow margins. Rather than form pups, this agave forms "bulbils," flowers that morph into little plantlets along the flower stalk. When the stalk falls over, the plantlets scatter over the ground and many root. Grow in full sun.

Aloe

Aloes are a diverse group of succulents mostly from South Africa, Madagascar, Jordan, and the Arabian Peninsula. Aloes can be as small as a tennis ball or as large as a tree. Some smaller aloes grow low to the ground and make great groundcovers or edgings. Others form wide rosettes that are right at home in the middle of a border. Shrublike aloes make great screens and backgrounds, while tree aloes provide the height that few other succulents have to offer. These flowering beauties are especially notable in winter, when some of the showiest species form enormous candelabras of coral, red, or orange flower clusters. The tube-shaped flowers are very attractive to hummingbirds. Aloes do great in the ground or in containers and they are adaptable to many garden styles. As a rule of thumb, grow aloes in full sun in California with minimal if any water once established. In Arizona, grow in filtered light or indirect bright light and water weekly in summer, and at most monthly in winter. A few aloes are cold hardy enough to grow outdoors year-round in New Mexico landscapes. Plant others in pots that can be moved to a protected area or indoors in fall when the weather cools.

Agave vilmoriniana 'Stained Glass'

Aloe 'Blue Elf'

COLD TOLERANCE 15°F
INTEREST winter to spring
SIZE 1–2 ft. tall, 1 ft. wide, eventually forming broad masses
COLOR blue-green, orange
SUCCULENT

'Blue Elf' is a small cold-hardy aloe with blue-green blades edged in soft white teeth. The blades form upright vases, which over time multiply into clusters and form patches. Plants bloom most from winter to spring, creating broad masses of bloom. Flowers form as spikes of pendant, coral-colored tubes edged in yellow. A favorite of hummingbirds. Grow as an edging or at the base of large pots set into the landscape.

Aloe cameronii

COLD TOLERANCE 25°F
INTEREST winter to spring and year-round
SIZE 1–2 ft. tall, 2–3 ft. wide, forming masses
COLOR blue-green, orange, brick red
SUCCULENT

Aloe cameronii produces succulent rosettes with wavy blades reminiscent of a starfish. The blades turn a striking brick red when planted in full sun. Plants sucker to make broad masses. Spikes of orange flowers appear in winter. Hummingbird favorite. Grow in full sun. When grown in too much shade or watered too generously, blades turn green.

Aloe 'Blue Elf'

Aloe cameronii

Aloe 'David Verity'

COLD TOLERANCE 20–25°F
INTEREST winter and year-round
SIZE 6 ft. tall and wide
COLOR soft green, red, pale yellow, and fiery orange
SUCCULENT

This clumping aloe develops multi-branched flower spikes of tall, deep red buds in winter that turn pale yellow, then open, revealing golden orange stamens. The tricolor flower spikes top succulent rosettes of long, narrow green or blue-green blades. They are absolutely beautiful. The blades are toothed and slightly reddish along their edges. Plant in full sun or light shade.

Aloe ferox
cape aloe

COLD TOLERANCE 20–25°F
INTEREST fall to winter and year-round
SIZE 6–9 ft. tall, 4 ft. wide
COLOR deep green or blue-green, fiery red-orange
SUCCULENT

This stately tree aloe is native to South Africa. It forms a single rosette of deep green or blue-green succulent blades atop a corky trunk. Blades have wide spines along their edges and covering their surfaces, which creates an interesting texture. In fall or winter, plants form large candelabras of fiery red or orange flower clusters. This beautiful specimen plant attracts hummingbirds. Grow in full sun.

Aloe 'David Verity'

Aloe ferox

Aloe rubroviolacea
Arabian aloe

COLD TOLERANCE 20–25°F
INTEREST winter and year-round
SIZE 2–3 ft. tall, 3–4 ft. wide, forming patches
COLOR green with violet blush, coral
SUCCULENT

One of my favorite aloes, *Aloe rubroviolacea* has broad succulent rosettes that are teal blue–tinged rose or lavender. In winter, each rosette forms spikes of dusky to bright coral-orange flowers attractive to hummingbirds. Rosettes tend to face the sun, and they sucker to make wide patches. Grow in full sun.

Aloe speciosa
tilt-head aloe

COLD TOLERANCE 20–25°F
INTEREST fall and winter
SIZE 8–10 ft. tall, 4–6 ft. wide
COLOR blue-green, pink blush, coral, and creamy white
SUCCULENT

This beautiful specimen shrubby aloe has rosettes of long succulent blades that are bluish green with pinkish edges and sit on a slight tilt on the end of upright stalks. Tall flower spikes produce coral-colored buds that open to creamy white flowers in sequence so the inflorescences look like giant spirals. Grow in full sun.

Aloe rubroviolacea

Aloe speciosa

Aloe spicata

COLD TOLERANCE 25–30°F
INTEREST winter
SIZE 4–6 ft. tall and wide
COLOR golden yellow, green
SUCCULENT

This African native is a golden yellow–flowering aloe that blooms in winter from rosettes of long, green succulent blades that turn coppery orange in dry situations. Plants form tall corky trunks. Plant in well-drained soil in light shade or full sun.

Aloe spicata

Alstroemeria 'Third Harmonic'
Peruvian lily

COLD TOLERANCE 15–20°F
INTEREST spring and summer
SIZE 2–3 ft. tall, 4 ft. wide
COLOR bright green, bright orange, yellow, and burgundy
PERENNIAL

This flowering perennial Peruvian lily forms patches of green, upright stems that are topped in spring and summer with clusters of bell-shaped, apricot-orange flowers with blush yellow throats and burgundy dots and dashes. These are excellent as cut flowers, but be sure to yank—don't cut—flower stalks from the

Alstroemeria 'Third Harmonic'

base to stimulate more flower production. Grow in full sun in mild climates, or some shade in hotter areas. Peruvian lilies are native to Chile and Brazil, but most plants on the market are hybrids. Look for their many different color combinations, from white to almost purple, salmon, coral, pink, red, orange, burgundy, and yellow. Each has a different variety name.

Anigozanthos
kangaroo paw

Kangaroo paws are upright perennials with clumping fans of bright green, irislike leaves. From the center of the fans, tall flower spikes arise topped with sprays of furry colorful buds. As buds open, they form long tubes that flare at the mouths, revealing pale greenish throats and a flower shaped like a tiny kangaroo paw. Cut faded flower stalks to the base so plants rebloom. Groom to remove dead leaves on occasion. If after a few years flowering diminishes, dig up and divide the plants. Some plants live only four or five years, yet are so stunning that they are more than worth planting and replanting. Kangaroo paws are excellent design accents, and do well in containers. They are fabulous when massed, great for cut flowers, and beloved by hummingbirds. Plants in this genus prefer well-drained soils and full sun to part shade.

Anigozanthos 'Big Red'
big red kangaroo paw

COLD TOLERANCE 20–25°F
INTEREST spring to fall
SIZE 4–6 ft. tall, 2–3 ft. wide
COLOR bright green, deep red, pale green
PERENNIAL

This tall kangaroo paw forms sprays of furry deep red buds, which open with pale greenish throats. A long-used and reliable variety.

Anigozanthos 'Big Red'

Anigozanthos 'Bush Dawn'

bush dawn kangaroo paw

COLD TOLERANCE 25–30°F
INTEREST spring to fall
SIZE 3 ft. tall and wide
COLOR bright green, bright yellow, with pale green
PERENNIAL

'Bush Dawn' forms sprays of furry bright yellow buds that open with pale greenish throats. The Bush Gem hybrid series are resistant to inkspot (a fungus that causes black spots on the leaves).

Arbutus 'Marina'

strawberry tree

COLD TOLERANCE 0°F
INTEREST year-round
SIZE 30–40 ft. tall and wide
COLOR cinnamon red, deep green, pink, green, yellow, red
TREE

This evergreen tree produces a rounded canopy of green, leathery leaves over shreddy, cinnamon red–colored bark. In spring, it is covered in pinkish flowers that look like tiny, upside down urns. These flowers become round, bumpy, marble-sized fruits that start out green, turn gold, and then strawberry colored. They are edible but not palatable for people; birds and other animals, however eat them, making them good habitat plants. Roots are not usually problematic for sidewalks or foundations. *Arbutus* is related to California's native manzanita (*Arctostaphylos*). 'Marina' tolerates many kinds of soils and typical garden conditions better than the California native *Arbutus menziesii* (madrone). Site in full sun. Available as a single- or multi-trunk tree. Fabulous as a background or screening plant.

Anigozanthos 'Bush Dawn'

Arbutus 'Marina'

Arctostaphylos

manzanita

Manzanitas are evergreen woody shrubs native to chaparrals habitats from western Canada down through California to Mexico. Some grow as far east as Texas. Most manzanitas, however, are native to California, from coast to mountains. As a group, manzanitas have twisted, mahogany- to burnt orange–colored bark (smooth or shreddy) and deep green, blue-green, or ice green leaves that can be reddish or coral-colored when young. Some have fuzzy silver leaves. Interestingly, the leaves often point upward, which helps keep them out of direct sunlight. Manzanitas' winter and early spring flowers form clusters of tiny, upside down urns in shades of white to cream to pink (and reportedly fragrant though I've never noticed a fragrance). Flowers turn to rusty orange or deep red berries (*manzanita* translates to "little apple") that support wildlife. Groundcover manzanitas grow only a foot tall and several feet across; others are shrub or small tree sized. Manzanitas are best grown in the ground, as opposed to in pots. All are beautiful, need little if any water when planted in the right spot, and require only minimal ongoing maintenance. Prune to reveal their spectacular sculptural shapes and stunning bark colors.

Arctostaphylos densiflora 'Sentinel'

sentinel manzanita

COLD TOLERANCE 15°F
INTEREST year-round
SIZE 4–8 ft. tall, 3–6 ft. wide
COLOR gray-green, pink, reddish brown
SHRUB

Sentinel manzanita is more upright than wide and features reddish-brown bark. Winter flowers are bright pink while leaves are a soft grayish green. Grow in full sun or part shade. Tolerates heavy soils but also does well in sandy soils. In wetter regions, there's no need to irrigate; where it is drier, irrigate up to once every four weeks in summer.

Arctostaphylos densiflora 'Sentinel'

Arctostaphylos densiflora 'Sentinel'

Arctostaphylos edmundsii var. *parvifolia*
bronze mat manzanita

COLD TOLERANCE 15°F
INTEREST year-round
SIZE 6 in. tall, 4–6 ft. wide
COLOR pink blush, red, green, brown
SHRUB

This diminutive manzanita is a perfect groundcover. Pink blush flowers appear in winter and early spring, turning to red berries. Leaves are glossy green. Plant in full sun along the coast, and part shade inland, into either well-draining or clay soils. Water infrequently in summer, if at all.

Arctostaphylos 'John Dourley'
John Dourley manzanita

COLD TOLERANCE 5–15°F
INTEREST year-round
SIZE 3–4 ft. tall, 6 ft. wide
COLOR pink, blue gray, bronze, red, brown
SHRUB

Pink flowers cover this lower, broader manzanita that grows into a beautiful blue-gray mound. New leaves start out bronze. Pale pink springtime flowers become soft red berries. Tolerates clay, full sun, and part shade. In wetter regions, there's no need to irrigate; where it is drier, irrigate once a month in summer.

Arctostaphylos edmundsii var. *parvifolia*

Arctostaphylos 'John Dourley'

Arctostaphylos 'Lester Rowntree'
Lester Rowntree manzanita

COLD TOLERANCE 15°F
INTEREST year-round
SIZE 8–10 ft. tall and wide
COLOR blue, green, pink, red, deep red
SHRUB

This manzanita tends to have a broad and graceful profile. It features bluish leaves and bright pink flowers starting in fall, eventually producing red berries. The lighter-colored leaves contrast beautifully with the smooth and gently twisted deep red bark. Grow in full sun or part shade. In wetter regions, there's no need to irrigate; where it is drier, irrigate once every two to four weeks in summer, depending on your location.

Arctostaphylos pungens
pointleaf manzanita

COLD TOLERANCE 15°F
INTEREST year-round
SIZE 3–10 ft. tall, 3–6 ft. wide
COLOR mahogany red, gray green, white
SHRUB

Native to the Southwest, from Texas to California to Utah and Mexico, this upright, vase-shaped shrub is taller than it is wide. Red-mahogany bark contrasts with gray-green leaves and white flowers. In its native habitat, the seeds sprout only following fire. Grows best in well-draining soils, but will tolerate clay soils.

Arctostaphylos 'Lester Rowntree'

Arctostaphylos pungens

Arctostaphylos 'Sunset'

sunset manzanita

COLD TOLERANCE 5–15°F
INTEREST year-round
SIZE 4–6 ft. tall and wide
COLOR copper, orange, gold, green, white, pink blush, mahogany
SHRUB

Sunset manzanitas new spring leaves emerge in the most beautiful colors: orange, copper, and gold, eventually maturing to green. Because it tolerates full sun to full shade, you could plant sunset manzanita under a madrone (*Arbutus*) to echo the colors of madrone bark. Sunset manzanita's bark is rich mahogany colored and shreddy. Winter and early spring flowers are white or pink blush. Tolerates clay soils and thrives in well-draining soils.

Arctostaphylos 'Sunset'

Arctotis

African daisy

Arctotis is a genus of evergreen perennials from South Africa with velvetlike, silvery-gray leaves that grow along horizontal stems. Plants grow into low, broad patches. In the short days of fall and spring, African daisies are covered in flowers in shades of white to yellow, orange to pink, and red to multicolored. Plants in this genus are easy to care for, and very dry growing. Plants flower best and stay denser in full sun, but tolerate some shade. Deadhead to prolong bloom. Irrigate deeply to establish, then only occasionally. When patches grow too leggy, wait until flowering is over, then cut stems back to about 6 inches long. Water deeply every four or five days until leaves resprout, then keep watering deeply but less and less frequently until plants fill back in, at which point return to only occasional irrigation. Root cuttings of nonpatented varieties (it is illegal to root patented varieties) in well-draining potting soil. African daisies are wonderful as front-of-border edging plants or under and around shrubs and trees. They are a beautiful complement to succulents as well and are virtually pest free.

Arctotis The Ravers 'Pink Sugar'
pink sugar African daisy

COLD TOLERANCE 20–25°F
INTEREST spring and fall
SIZE less than 1 ft. tall, 2–4 ft. wide
COLOR pink, sherbet orange, gray-green
PERENNIAL

Pink sugar African daisy produces flowers, 3–4 inches in diameter, that are pale orange in the center and deep pink toward the edges. The contrasting silver-gray foliage enhances this unusual, eye-catching color combination.

Arctotis The Ravers 'Pumpkin Pie'
pumpkin pie African daisy

COLD TOLERANCE 25–30°F
INTEREST spring and fall
SIZE 1 ft. tall, 2–4 ft. wide
COLOR gray green and pumpkin orange
PERENNIAL

Deep pumpkin orange flowers, 3–4 inches in diameter, against green-gray foliage define this African daisy.

Arctotis The Ravers 'Pink Sugar'

Arctotis The Ravers 'Pumpkin Pie'

Berlandiera lyrata
chocolate flower

COLD TOLERANCE −30°F
INTEREST late spring to autumn
SIZE 1–2 ft. mounds
COLOR green, yellow
PERENNIAL

Chocolate flower is a Southwest native, with daisy-like, quarter-sized yellow flowers in summer and fall. Yellow "petals" are brown striped on the undersides and have burgundy-brown centers (where the actual tiny flowers sit). Blooms release their chocolate scent at night. These plants have sculpted gray-green foliage and are desert perennials. Grow in full sun or part shade, especially in desert areas. Deadhead to prolong bloom. Along the coast, chocolate flowers are most often grown as annuals. Attractive to butterflies.

Beschorneria yuccoides
Mexican lily

COLD TOLERANCE 15°F or cooler
INTEREST year-round
SIZE 6 ft. tall and wide
COLOR gray green, coral pink, chartreuse
SUCCULENT

If you like the shape of agaves but want a softer look, this agave cousin fits the bill. *Beschorneria yuccoides* has long, softly succulent gray-green or green blades arranged in a rosette. Blades are pliable and smooth, without teeth or spines. They can also be slightly curved. Once mature, a rosette can send up a tall (8 feet or taller), curved coral pink flower stalk with tiny chartreuse flowers. After the flower fades, the rosette dies, leaving behind many pups. Grow in full sun in California gardens, or some shade in low-desert gardens. Requires little if any water once established.

Berlandiera lyrata

Beschorneria yuccoides

Bignonia capreolata 'Tangerine Beauty'

cross vine

COLD TOLERANCE 10–15°F
INTEREST summer
SIZE 20–30 ft. long
COLOR orange, yellow, green
VINE

In spring and summer, this beautiful and vigorous evergreen vine has bright green leaves and exuberant coral to orange to brick red, trumpet-shaped, 2-inch-long flowers. It clings by tendrils as it climbs, so it needs strong support for long branches. Cross vine is native to the United States, from Florida to Ohio, Illinois to Texas. It tolerates infrequent irrigation once established in coastal areas; it is thirstier in desert areas but excellent for creating cooling shade when trained over a trellis, pergola, or up a wall. Plant in full sun. Hummingbirds cannot resist this plant.

Bougainvillea

Bougainvilleas are thorny, woody vines from South America that bring bright colors to hot, dry gardens. When most people think of bougainvillea flowers, they think of the bright-colored, papery bracts; the actual flowers are small, tubular structures hidden at the center of the bracts. Either way, the vivid color of the genus is what people clamor for: shades of red, pink, coral, fuchsia, white, orange, or gold. Standard bougainvilleas grow 20–30 feet long and require substantial support structures (they need to be actively trained and tied to the supports since they don't attach themselves). Dwarf varieties range from 4 feet by 8 feet to upright plants that are just 2–3 feet tall and wide. Most have green leaves, though some have variegated green and pale yellow leaves. Leaves overwinter if temperatures remain at or above freezing. In colder regions plants can die back to the base in winter. Bougainvilleas are very thorny and best grown away from walkways or patios. If not properly pruned and trained, they can become dense thickets with bare wood in the middle. Their constant rain of bracts and leaves requires a lot of maintenance. When grown as groundcovers or in garden beds, the debris acts as a mulch, though plants still need to be trained. Plant in full sun, and treat with tough love. Too much water or fertilizer inhibits blooms.

Bignonia capreolata 'Tangerine Beauty'

Bougainvillea 'Bengal Orange'

COLD TOLERANCE 30°F
INTEREST spring to fall
SIZE 4 ft. tall, 8 ft. wide
COLOR creamy yellow, green, coral, pink
VINE

This smaller bougainvillea has cream-edged green leaves that are pink blush when first emerging. Flowers open coral and age to pink. The multiple color combinations are especially eye catching and combine well with pink-, yellow-, and orange-blooming plants, and plants with blue foliage. In colder climates, grow in a pot and move to a protected location when temperatures dip below freezing.

Bougainvillea brasiliensis

COLD TOLERANCE 20–25°F (defoliates at 30°F)
INTEREST spring to fall
SIZE stems 20–30 ft. long
COLOR pink-purple, green
VINE

This classic bougainvillea has arching stems, bright green leaves, and masses of brilliant purple or pink flowers.

Bougainvillea 'Bengal Orange'

Bougainvillea brasiliensis

Brahea armata
blue hesper palm, Mexican fan palm

COLD TOLERANCE 15°F
INTEREST spring and year-round
SIZE 15–40 ft. tall, 20 ft. wide
COLOR silver blue, creamy yellow
PALM

This single-trunk, silver-blue-leaved fan palm is native to Baja California's low deserts and canyons. It is dry and slow growing along the coast, and grows faster and taller in the desert. In spring, blue hesper palm develops spectacular 15-foot-long, arching sprays of creamy green-yellow flowers. Does best in full sun and well-draining soil, needs little or no irrigation once established, but grows faster and will look a bit better with monthly irrigation in summer's heat.

Buddleia marrubifolia
wooly butterfly bush

COLD TOLERANCE 10°F
INTEREST spring to summer
SIZE 3–5 ft. tall and wide
COLOR silver and golden orange, gold
SHRUB

This unusual and beautiful butterfly bush features velvety pale silver leaves that contrast beautifully with marble-sized balls of tiny golden orange flowers that top its branches most of the year. This shrub prefers well-drained soils and tolerates intense heat and sunlight with little to no irrigation. Along the coast, if you water it too much, it will grows quickly and then burn out. In desert areas, irrigate deeply on occasion. Prune periodically to maintain density and structure. Visited frequently by butterflies.

Brahea armata

Buddleia marrubifolia

Calliandra 'Sierra Starr'
Sierra starr fairy duster

COLD TOLERANCE 15°F
INTEREST year-round
SIZE 5 ft. tall and wide
COLOR watermelon red, green
SHRUB

Cross a pink fairy duster with the Baja fairy duster and this is what you get: a lovely shrub covered most of the year in feathery green leaves and tufts of bright watermelon red flowers that attract hummingbirds. This desert legume grows in full sun in hot climates. If shrub grows too dense, prune occasionally, but only after flowering.

Calliandra 'Sierra Starr'

Calylophus drummondianus
sundrop

COLD TOLERANCE 0°F
INTEREST spring to fall, depending on location
SIZE 1 ft. tall, 1–2 ft. wide
COLOR yellow, green
PERENNIAL

This diminutive shrubby perennial is blanketed in quarter-sized bright yellow blooms nearly year-round in California gardens. In colder climates, flowers appear in spring and summer. Narrow, bright green leaves clothe the branches. Mass along the edges of beds or surrounding trunks of aloes, madrones (*Arbutus*), or large agaves. It can also be grown in containers. Sundrop is native from Louisiana to New Mexico and north through Colorado, where it grows in sandy, gravely soils (this tells you that the plant does best in well-draining soils). This is one tough plant; it will thrive in full sun and with little if any irrigation once established. Cut back in spring to rejuvenate. Also known as *Calylophus berlandieri* subsp. *berlandieri*, *Calylophus drummondii*.

Calylophus drummondianus

Cassia

Cassia is a large genus of woody shrubs and trees in the legume family. Those featured here are all Australian natives from hot, tough, dry climates. Each has short, narrow phyllodes, which look like leaves. From fall through early spring, and occasionally throughout the year, they sport clusters of pealike bright yellow flowers that form bright green pods that turn coppery brown and all hang on the plant at once. The color combination is quite striking, as is the architecture of the shrubs. Plants in the genus are easy care; grow in full sun or in the dry shade of high-canopy trees. They prefer well-draining soils. In California, they are very drought tolerant; water deeply, but only occasionally, in Arizona summers.

Grow these with the deep green–leaved and blue-flowered California lilacs (*Ceanothus*) for a fantastic color and shape contrast in California gardens. Also known as *Senna*.

Cassia artemesoides
feathery cassia

COLD TOLERANCE 15°F
INTEREST fall to spring
SIZE 6 ft. tall and wide
COLOR silver-green, yellow, chartreuse, copper
SHRUB

Feathery cassias have silvery green, needlelike blades arranged geometrically along branches.

Cassia artemesoides

Cassia nemophila
desert cassia

COLD TOLERANCE 10°F
INTEREST fall to spring
SIZE 6–8 ft. tall and wide
COLOR olive green, yellow, chartreuse, copper
SHRUB

Desert cassia features olive green to bright green blades that are narrow and needlelike. This cassia is more cold tolerant than *Cassia artemesoides* or *C. phyllodinea*.

Cassia phyllodinea
silvery cassia

COLD TOLERANCE 20–25°F
INTEREST fall to spring
SIZE 6 ft. tall and wide
COLOR soft green, yellow, chartreuse, copper
SHRUB

These shrubs form gray-green, sickle-shaped blades, with yellow flowers and chartreuse seed pods that turn copper.

Cassia nemophila

Cassia phyllodinea

Ceanothus
California lilac

California lilacs have an unfortunate common name since they are nothing at all like lilacs. Instead, they are a diverse genus of 50 or 60 evergreen shrubs hailing from Canada to Florida to Guatemala. Of those, 41 species are native to the California floristic province, from the coast to the tallest mountains. Mediterranean-climate California lilacs are evergreen shrubs with woody branches covered with leaves from 3 inches long and about half as wide, to tiny, rounded leaves, shaped like the head of a pencil eraser. Many species have tough, leathery leaves; some are deeply grooved, smooth, or even slightly hairy.

As a rule, California lilacs are adapted to soils with minimal organic matter and low fertility—another characteristic typical of Mediterranean climates. Springtime flower clusters range from white to deep blue, almost purple, and are very attractive to wildlife. Plants in the genus are also nitrogen fixers, so they don't need fertilizer. Plant unamended in native soils. In clay soils, plant into large mounds of well-draining soil. Water to establish, then stop. Death from overwatering is one of the reasons some varieties have a reputation for being short lived. These are excellent plants for wildlife. Rarely grown outside California.

Ceanothus 'Centennial'
centennial California lilac

COLD TOLERANCE 15–20°F
INTEREST spring
SIZE less than 1 ft. tall, 4–6 ft. wide
COLOR deep green, cobalt blue
SHRUB

This low-growing groundcover's leaves are deep green and grooved. Its flowers are deep cobalt blue. Grow in sun, or part sun in hotter regions. Needs minimal irrigation after establishment along the coast, but more inland. This is a great choice for a slope or as a negative green space to replace unused lawn (as long as you don't need to walk on it). Fast growing, and prefers well-draining soils.

Ceanothus 'Centennial'

Ceanothus 'Concha'

concha California lilac

COLD TOLERANCE 10°F
INTEREST spring
SIZE 6–8 ft. tall, 8–12 ft. wide
COLOR deep green, intense blue
SHRUB

'Concha' is a popular garden cultivar with rounded clusters of beautiful, deep blue flowers. Branches are slightly arched and densely covered in narrow, deep green leaves. Fairly tolerant of both heavy soils and overwatering. Grow in full sun.

Ceanothus 'Dark Star'

dark star California lilac

COLD TOLERANCE 15°F
INTEREST early spring
SIZE 4–8 ft. tall, 8–12 ft. wide
COLOR deep green, purple, vibrant blue
SHRUB

This popular cultivar is especially suited for coastal gardens. Leaves are small and deep green. In early spring, this shrub erupts in grape bubblegum purple buds, followed by vibrant blue flowers. Fast growing and upright, it is perfect for an informal hedge or screen. Grow in full sun.

Ceanothus 'Concha'

Ceanothus 'Dark Star'

Ceanothus 'Frosty Blue'
frosty blue California lilac

COLD TOLERANCE 15°F
INTEREST spring
SIZE 6–10 ft. tall, 8 ft. wide
COLOR dark green, frosty blue and white
SHRUB

Upright 'Frosty Blue' grows quickly. It can be grown as a garden hedge, or pruned to more of a tree shape. Leaves are dark green, while blue flowers show bits of white that suggest "frosting." Tolerates heavy soils but prefers well-draining soils. Water deeply on occasion through summer.

Ceanothus gloriosus var. exaltatus 'Emily Brown'
Emily Brown California lilac

COLD TOLERANCE 10°F
INTEREST winter and spring
SIZE 3–4 ft. tall, 8–10 ft. wide
COLOR dark green, dark violet blue
SHRUB

This groundcover has long, arching branches covered in dark green leaves. Clusters of late-winter and early-spring flowers, 1 inch across, are dark violet blue. Plants work well in slope plantings along the coast. Tolerates clay soils.

Ceanothus 'Frosty Blue'

Ceanothus gloriosus var. exaltatus 'Emily Brown'

Cistus

rockrose

Rockroses are Mediterranean-native evergreen shrubs with hairy, sometimes sticky green leaves. These shrubs bloom prolifically in spring and sometimes into summer with simple, broad rose-like flowers, typically with five crepe papery petals. The flowers, which open and drop in the same day, tend to be in shades of pink or white, some with an attractive reddish spot at the base of each petal, just outside the fringed yellow center. These drought-tolerant shrubs thrive in poor, well-draining soils, respond to tough love, and seem to fail when overwatered. I always see the healthiest specimens growing on slopes. Rockroses are primarily grown in California but I've seen them in protected areas of gardens in Albuquerque, too.

Cistus ×*purpureus*

purple rockrose

COLD TOLERANCE 15°F
INTEREST spring and summer
SIZE 4–6 ft. tall, 5–7 ft. wide
COLOR olive green, intense pink, red
SHRUB

This rockrose has deep pink (not purple) flowers, 3 inches in diameter, with a red blotch at the base of each petal. This mounding shrub has narrow, slightly undulating leaves that are dark green on top and dusky green underneath and resinous. Tolerates coastal salt spray as well as milder desert conditions.

Cistus ×*purpureus*

Cistus ×*skanbergii*
pink rockrose

COLD TOLERANCE 20–25°F
INTEREST spring and summer
SIZE 3–4 ft. tall, 4–5 ft. wide
COLOR pale pink, yellow, gray green
SHRUB

Dense, small, grayish-green leaves cover this mounding pink rockrose shrub. Spring and summer flowers are pale pink with a yellow center, each just an inch across. Tolerates salt spray as well as milder desert conditions.

Cistus ×*skanbergii*

Clivia
orange clivia

COLD TOLERANCE 20–25°F
INTEREST fall and spring, or year-round, depending on location
SIZE 1–2 ft. tall and wide
COLOR deep green, bright orange, sometimes buttery yellow
BULB OR RHIZOME

Like irises, clivias are fleshy rooted perennials often promoted as the go-to bulbs to light up dry shade in California gardens. They are South African natives with tongue-shaped, deep glossy green leaves that curve gracefully and form wide fans. Bright orange bugle-shaped flowers form in clusters atop stalks that rise from the fans and bloom in the cool seasons of the year. Look for the rarer yellow-blooming cultivars too. Clivias ask only for well-drained soil and occasional irrigation. I've seen decades-old plantings growing with absolutely no irrigation in shaded coastal and valley gardens. Tuck into narrow garden beds under eaves, around the base of a newly planted tree, or grow in a container. Grown as a houseplant in desert regions.

Clivia miniata

Cordyline

Cordylines are tall, narrow plants with slender, upright stems more corky than woody, awith columns of strappy, narrow leaves, more pliable but similar in appearance to *Yucca* or *Dracaena*. Young plants tend to have a single stem and head, but as plants mature, they develop multiple heads, each atop its own branch. Typical garden cordylines are New Zealand natives (or hybrids of New Zealand natives) grown for their colorful foliage rather than for blooms, which are occasional, subtle, and comprised of surprisingly fragrant tiny flowers that develop along long stalks. Many recent introductions, grown as alternatives to *Phormium* (New Zealand flax), are bred or selected to be "trunkless" so they remain as short plants. Cordylines are popular plants in California gardens, where they contribute their height, texture, and color without width or shade. They are great container plants.

Cordyline australis 'Torbay Dazzler'

grass tree

COLD TOLERANCE 10–15°F
INTEREST year-round
SIZE 25 ft. tall, 3 ft. wide
COLOR yellow cream, dusky green, red blush
GRASSLIKE

This cordyline grows slowly and stays narrow, with a head just 3 feet across. As it matures, it forms multiple heads. Blades are beautifully variegated, striped yellow cream along the outer margins, dusky green in the center, with a red blush at the base of each leaf and a narrow reddish stripe. Grow in full sun in coastal California, or in some shade farther inland.

Cordyline australis 'Torbay Dazzler'

Cotinus coggygria
smoke tree

COLD TOLERANCE –25°F
INTEREST spring to late fall
SIZE 10–15 ft. tall and wide
COLOR burgundy-purple, blue-green, apricot, vermillion
TREE

This deciduous plant can be either a large shrub or a small tree. Smoke tree is native to parts of Asia and Europe but is grown around the world. The name "smoke tree" comes from the airy filamentous flower clusters that develop in summer. They are interesting, but not as attractive as the round, deep burgundy or purple leaves that turn apricot and vermillion in fall before they drop. Some cultivars have new leaves that emerge purple, then mature a deep blue-green. Look for 'Purple Robe' or 'Royal Purple', which have the deepest plum-colored leaves. Plant in full sun or light shade, especially in desert climates. Provide little to no irrigation once established in mild climates, more in desert climates.

Cotinus coggygria

Cordyline 'Design-A-Line Burgundy'

burgundy cordyline

COLD TOLERANCE 10°F
INTEREST year-round
SIZE 3 ft. tall and wide
COLOR burgundy
GRASSLIKE

'Design-a-Line Burgundy' is a trunkless cordyline whose dense clusters of slightly curved blades are burgundy colored. It is drought tolerant along the coast, thirstier inland. Grow in full sun in coastal California, in some shade farther inland.

Crassula

Crassula is a diverse group of succulents, mostly from South Africa. These are smaller-scale plants, many with brightly colored foliage, and many with interesting, even architectural foliage. They make tiny white, star-shaped blooms, but they are mostly grown for foliage instead of blooms. Plants in this genus are extremely easy to propagate from cuttings or even leaves. Grow in the ground in well-draining soil, though they make excellent container plants, too.

Cordyline 'Design-A-Line Burgundy'

Crassula capitella 'Campfire'

COLD TOLERANCE 30°F
INTEREST year-round
SIZE 6 in. tall, 2–3 ft. wide
COLOR Gold, orange, red, green
SUCCULENT

This low, shrubby succulent has foliage that turns fiery shades of chartreuse to gold, sherbet orange to red rust, especially in the cool of winter. Near the coast, provide full sun, part sun, or light shade and minimal irrigation. Inland, grow in part shade and water a bit more. Great as a groundcover and excellent in containers glazed oxblood red or bright green. Dormant in summer.

Crassula falcata
airplane plant, propeller plant

COLD TOLERANCE 20 to 25°F
INTEREST summer
SIZE 2–3 ft. tall and wide
COLOR blue-green, scarlet
SUCCULENT

Airplane plant has sword-shaped, blue-green blades that swirl around the center stalk like a twirling helicopter rotor. Summer flowers are tiny, clustered together atop a stalk to resemble scarlet broccoli—an odd and amazing look. It is very architectural, and has good texture and great color. Plants grow in full sun or part sun with just occasional summer irrigation. Does best if kept dry in winter.

Crassula capitella 'Campfire'

Crassula falcata

Crassula ovata 'Hummel's Sunset'
Hummel's jade plant

COLD TOLERANCE 30°F
INTEREST year-round
SIZE 2–3 ft. tall and wide
COLOR golden yellow, amber, red, bright green
SUCCULENT

This selection of the classic jade plant is a bit on the small side, with upright, corky succulent stems covered with rounded, glossy leaves in shades of golden yellow, bright green, amber, and red in the cooler months. Grow in full sun with little irrigation. Does well in the ground or in a container—it can even be grown as a bonsai. Too much shade, fertilizer, or water makes for greener leaves.

Dalea capitata
Sierra gold dalea

COLD TOLERANCE 0°F
INTEREST spring and fall
SIZE less than 1 ft. tall, 3–4 ft. wide
COLOR yellow, green
GROUNDCOVER

For a pretty and petite-looking groundcover, look to this dalea, which has fragrant, fine, fringy green foliage and short flower stalks topped in tiny towers of yellow flowers in spring and fall. This desert native also makes an excellent edging plant. It is fast growing in full sun, intense heat, and well-drained soil with little irrigation. It may lose leaves in cold winters. *Dalea capitata* is better suited to desert regions than coastal areas.

Crassula ovata 'Hummel's Sunset'

Dalea capitata

Echeveria

Echeverias are rosette-forming succulents, typically with blue, teal, silver aqua blue-green, green, or ice green foliage, and pinkish flowers that form on long spikes. The rosettes make them much-prized garden plants, both in the ground and in containers. Some varieties develop unattractive, tall stalks over time. To reroot, simply cut off the head along with a few inches of stalk and replant. Echeverias are native to Central and South America. Grow in well-draining soils and bright light or full sun. Pull off spent leaves. Divide offsets to multiply.

Echeveria 'Afterglow'

COLD TOLERANCE 20–25°F
INTEREST year-round
SIZE 1–2 ft. tall and wide
COLOR pink, dusky mauve, coral
SUCCULENT

'Afterglow' produces rosettes of smooth-surfaced leaves that emerge icy aqua blue, then turn pink and dusty mauve. Growing this plant in a pot helps to show it off and keep it from being bruised or damaged.

Echeveria 'Afterglow'

Echeveria agavoides 'Lipstick'
lipstick echeveria

COLD TOLERANCE 15–20°F
INTEREST year-round
SIZE 6 in. tall, 1 ft. wide
COLOR bright green, bright red
SUCCULENT

This tight rosette-forming echeveria features pointed (but not sharp) yellow-green blades edged with roughly drawn bright red "outlines," that create a very dramatic and geometric effect. Rosettes are no more than a foot across and 8–10 inches high. These make beautiful single specimens in a container, though the plants eventually form side pups. Spring and summer flowers are red and yellow. Grow in full sun along the coast, or light shade inland.

Echeveria elegans
Mexican snowball

COLD TOLERANCE 20–25°F
INTEREST year-round
SIZE 8 in. tall and wide, forming colonies
COLOR aqua-silver, coral pink
SUCCULENT

Mexican snowball forms small, tight rosettes of aqua-silver blades. One plant quickly grows into a low, spreading colony that covers the top of a pot, fills spaces between boulders, and so on. In late winter and spring yellow flowers develop on tall coral pink stalks. Mexican native. Grow in full sun along the coast, light shade inland.

Echeveria agavoides 'Lipstick'

Echeveria elegans

Echeveria 'Etna'

COLD TOLERANCE 30°F
INTEREST year-round
SIZE 2 ft. tall and wide
COLOR purple, teal, green
SUCCULENT

'Etna' is one of the carunculated echeverias (carunculations are fleshy growths that create the rough, bumpy surfaces that develop on mature echeveria leaves). This one has the look of a violent eruption, which is why it is named for Sicily's Mount Etna volcano. Rather than spewing lava, this succulent spews fleshy blades that are purple, teal, red, and green. Rosettes grow to 2 feet across, each atop its own thick stem. Plant in full sun to part sun for best color.

Echeveria gigantea
giant hen and chicks

COLD TOLERANCE 20–25°F
INTEREST year-round
SIZE 1 ft. tall, 1–2 ft. wide rosettes
COLOR dusky teal, dusky pink
SUCCULENT

This plant's slightly wavy dusky teal blades are edged in muted rose and make an open, single rosette atop a corky succulent stem. In fall and winter, coral flowers with teal calyces cover dusky pink flower stalks that reach 3–6 feet tall. Grow in well-draining soil in the ground or in pots. Giant hen and chicks prefers full sun along the coast, or light shade inland.

Echeveria 'Etna'

Echeveria gigantea

Echeveria ×imbricata

COLD TOLERANCE 20–25°F
INTEREST year-round
SIZE 4–8 in. tall, 8 in. wide, forming colonies
COLOR blue, pink, yellow, coral
SUCCULENT

This is a classic succulent with round, blue rosettes touched with pink. Rosettes are only 4-6 or 8 inches across and somewhat flat topped. Single plants quickly become a colony of dozens, making this a great plant to site near rocks or boulders. Flowers are yellow on coral stalks and from coral buds in spring and summer. Plant in full sun. Also sold as Echeveria 'Imbricata'.

Echinocactus grusonii
golden barrel cactus

COLD TOLERANCE 15°F
INTEREST spines year-round, flowers spring to fall
SIZE 3–4 ft. tall, 8 ft. wide
COLOR golden yellow, green, bright yellow, white
CACTUS

Golden barrels are many-ribbed, globe-shaped bright green cacti covered in a gorgeous veil of golden yellow spines. These slow-growing cacti develop a patch of white "wool" at the top, where flowers emerge even brighter yellow than the spines. While most often single, older barrels sometimes pup to form clusters of seven or eight. These are stunning in a mass planting, especially when back-lit so that the spines glow in the late afternoon sun. Plant in full sun and well-draining soil. Golden barrels require very little if any water in desert gardens, and none in coastal or valley California gardens.

Echeveria ×imbricata

Echinocactus grusonii

Epilobium canum
California fuchsia

COLD TOLERANCE 15°F or cooler
INTEREST late summer and fall
SIZE 1–2 ft. tall, 3 ft. wide
COLOR frosty gray-green, brilliant red-orange
PERENNIAL

This low-growing, mat-forming California native perennial has frosty gray-green leaves. In late summer, the branches erupt with flowers the same shade of red-orange as pomegranate flowers. Bloom continues into fall, during which the brilliant red-orange against the frosty gray-green is a showstopper. The trumpet-shaped flowers are hummingbird magnets. Plants can go deciduous in the coldest months of the year, but resprout in spring. This is one of the most versatile and easy to grow of the California fuchsias. Grow in full sun and well-draining soils. Drought tolerant once established. Also known as *Epilobium californica*, *Zauschneria canum*, *Zauschneria californica*. 'Schieffelin's Choice' is a hybrid that grows only 6 inches tall.

Eriogonum
buckwheat

Eriogonum is a genus of mounding or mat-growing perennials, annuals, and shrublets that grow 3 feet tall by 3–5 feet wide. Flower stalks rise above the foliage to create a veil of flower clusters covering the entire mound. These are favorites of pollinators like butterflies, bees, birds, and small mammals. Most garden selections are California natives.

Epilobium canum

Eriogonum fasciculatum
California buckwheat

COLD TOLERANCE 10°F
INTEREST summer
SIZE 3 ft. tall, 5 ft. wide
COLOR white, green, rust
PERENNIAL

Eriogonum fasciculatum produces flowers that are white at first, eventually fading to an attractive rust color. This California chaparral and coastal sage scrub perennial prefers full sun and no water once established in California. *Eriogonum fasciculatum* var. *polifolium* is the best choice for desert gardens.

Eriogonum grande var. *rubescens*
red buckwheat, San Miguel island buckwheat

COLD TOLERANCE 15–20°F
INTEREST spring to summer
SIZE 1–2 ft. tall, 3–5 ft. wide
COLOR blue-green, white, rose
PERENNIAL

This California native evergreen shrublet has spatula-shaped, blue-green leaves that are white on the undersides. From late spring through summer, plants are veiled in brilliant rose-colored flower clusters. Cut back by a third after flowering if plant starts to look ragged. Tolerates clay soils. Prefers full sun near the coast, or some shade in hotter inland areas.

Eriogonum fasciculatum

Eriogonum grande var. *rubescens*

Eriogonum nudum 'Ella Nelson's Yellow'

Ella Nelson's naked buckwheat

COLD TOLERANCE −10°F
INTEREST spring to summer
SIZE 3 ft. tall, 2 ft. wide
COLOR golden yellow, green
PERENNIAL

While naked buckwheat typically grows from the coast to the mountains, this particular selection was found in Mendocino County, California, and seems to be more popular in the northern, higher rainfall regions of the state. What makes it distinctive is the buttons of beautiful yellow spring and summer flowers (rather than white or pink) that top gray-green leaves that are white underneath. Flowers fade to buff in fall. Prefers well-drained soil and full sun along the coast and in the northern parts of California, or some shade further south.

Eriogonum wrightii

Wright's buckwheat

COLD TOLERANCE −10°F
INTEREST fall
SIZE 18 in. tall, 2 ft. wide
COLOR silver-gray, white, pink
PERENNIAL

This arid-climate native has small gray leaves and white to pink flowers that top slender flower stalks from summer to fall in desert regions. Wright's buckwheat forms a flat, silvery mat. Prefers full sun. Very cold tolerant.

Eriogonum nudum 'Ella Nelson's Yellow'

Eriogonum wrightii

Eschscholzia

poppies

California is known as the Golden State, in part because of the gold rush and in part because of these golden orange wildflowers that grow from coast to mountain. These poppies are best planted from seed, during rainy winter months. Seeds germinate with rains, developing mounds of fine, feathery foliage in pale green to blue-green. As spring warms, flower stalks rise from the center. Their buds open to cup-shaped flowers with the typical (for a poppy) four petals, each satiny orange-gold, bright yellow, or pale yellow. After flowering, let plants reseed for the following year. Poppies prefer full sun and well-draining soil. No irrigation is required, but flowering lasts longer with occasional irrigation. These are very important plants for pollinators.

Eschscholzia californica

California poppy

COLD TOLERANCE −5°F
INTEREST spring
SIZE 12–18 in. mounds
COLOR silver, blue-green, golden orange
ANNUAL

California poppies are perennials grown as annuals throughout California and the southwest. They form feathery mounds with a deep taproot. Cultivars are red, white, pink, or yellow, but the golden orange straight species is still the most beautiful, with flowers that light up the early spring garden. Though they are perennials, California poppies are typically grown as annuals.

Eschscholzia californica

Eschscholzia californica subsp. *mexicana*
Mexican gold poppy

COLD TOLERANCE 0°F
INTEREST spring
SIZE 6 in. tall and wide
COLOR golden orange, green
ANNUAL

Mexican gold poppy is native from California to Texas. It grows more compact than the straight species but has no less impressive golden orange blooms that develop as the foliage develops.

Eschscholzia californica subsp. *mexicana*

Eucalyptus macrocarpa
mottlecah

COLD TOLERANCE 20°F
INTEREST spring and year-round
SIZE 8 ft. tall, 10 ft. wide
COLOR silver-white, brilliant red
SHRUB

Invasive, brittle-branched, and shallow-rooted, *Eucalyptus* trees have cast a pallor on the entire genus, which is a shame since there are many *Eucalyptus* trees and shrubs that make wonderful garden plants. *Eucalyptus macrocarpa* is one of the best shrubs for hot, dry climates. Its long, sprawling branches are densely clothed in broad, ghostly white, leathery leaves that are actually green with a thick waxy coating. In spring, the branches produce large, round, pointed-top white capsules that open to reveal brilliant red-fringed "flowers," easily 4 inches in diameter. Grow mottlecah in bright light and well-draining soil, with no irrigation in summer. This is a challenging plant to grow but entirely worth it if you can keep it going. Place a teal blue or cobalt glazed pot (either planted or left empty as architecture) next to this plant's branches—the colors are stunning together.

Eucalyptus macrocarpa

Euphorbia

Euphorbia is a diverse group of succulent and nonsucculent plants that range from tree sized to groundcovers. Most are grown for their colorful, textural shapes and foliage. Those that have colorful "flowers" are icing on the cake. The showy parts that look like flowers are actually colored bracts; the true flowers are tiny structures at the center of the bracts. Poinsettias (*Euphorbia pulcherrima*), for example, naturally have large colorful bracts that are bred for even more spectacular size and color. All euphorbias have irritating, sometimes toxic latex sap, so wear long sleeves, long pants, eye protection, and gloves to work with these plants.

Euphorbia antisyphilitica
candellia

COLD TOLERANCE 5°F
INTEREST late winter to fall
SIZE 1–2 ft. tall, 2–3 ft. wide
COLOR blue-green, red, pink, white
SUCCULENT

Euphorbia antisyphilitica has upright, pencil-thin, blue-green succulent stems dotted with tiny pink and white flowers (but no leaves) from late winter through fall. The stems spread slowly over time. The name candellia means "little candle," named after the plant's waxy coating, which helps prevent water loss. Plant in well-drained soil in very hot exposure, even in reflective heat. In desert areas, irrigate once a month, less often in cooler climates. The sap can be irritating, so wear protective clothing, glasses, and gloves when you work with the plant. Good in the ground, excellent in containers.

Euphorbia antisyphilitica

Euphorbia 'Blackbird'
blackbird spurge

COLD TOLERANCE 0°F
INTEREST spring and year-round
SIZE 1½–2 ft. tall and wide
COLOR purple-black
PERENNIAL

Blackbird spurge is renowned for its deep purple, almost black leaves, and red stems that contrast with its chartreuse flower bracts in spring. Tolerates full sun and occasional water.

Euphorbia characias
variegated spurge

COLD TOLERANCE 0°F
INTEREST spring and year-round
SIZE 3–4 ft. tall, 2 ft. wide
COLOR chartreuse, blue-green, variegated cream and green
SHRUB

Upright evergreen shrubs whose multiple branches emerge from a central point. Leaves are lance-shaped (but not sharp) at the tips. In spring, branches are topped in clusters of flowerlike bracts, bright chartreuse green in some common cultivars. Seedpods replace flowers and as seedpods dry, they burst open with a distinct "pop pop!" Over time, seeds spread through the garden so spring brings a lovely carpet of plants that unify the growing beds. Those that sprout in the "wrong" place are easily pulled, but do wear protection since sap is irritating and potentially toxic. Also wear protection to cut back brown branches at the end of the season but don't pull them out since they resprout at the base.

Euphorbia 'Blackbird'

Euphorbia characias 'Tasmanian Tiger'

Among common cultivars are 'Bruce's Dwarf' (green leaves and growing 2–3 feet tall and wide), 'Tasmanian Tiger' (variegated green and pale yellow leaves, growing 2–3 feet tall and wide), as well as *Euphorbia characias* subsp. *wulfenii*, which is all green and also grows 2–3 feet tall and wide.

Euphorbia cotinifolia
Caribbean copper plant

COLD TOLERANCE 30°F
INTEREST year-round
SIZE 20 ft. tall, 8–10 ft. wide
COLOR burgundy, gray
SHRUB OR SMALL TREE

Caribbean copper plant is a tree-sized euphorbia with corky gray bark and round cabernet-colored leaves that resemble the leaves of *Cotinus coggygria* (smoke tree). True flowers are cream colored, though they are tiny and barely noticeable. This plant is hardy and drought tolerant in only the most frost-free gardens, but in those gardens it is a spectacular specimen and easy to shape (wear protective clothing and glasses to avoid exposure to irritating sap). Excellent for protected courtyards and narrow spaces. Prefers full or part sun.

Euphorbia cotinifolia

Euphorbia × *Martinii* 'Ascot Rainbow'

ascot rainbow spurge

COLD TOLERANCE −10°F
INTEREST year-round
SIZE 2½ ft. tall, 2 ft. wide
COLOR lime green and butter yellow, blush red and deep red
PERENNIAL

Ascot rainbow spurge's upright branches emerge from a common base and are covered in variegated, lime green blades edged in butter yellow. New leaves emerge with a deep red blush that returns with the cool temperatures of fall. Chartreuse flower bracts with tiny deep red flowers in the center make the flowers appear to have red eyes. Drought tolerant once established.

Euphorbia milii

crown of thorns

COLD TOLERANCE 30°F
INTEREST year-round
SIZE 1–4 ft. tall, 1–3 ft. wide
COLOR bright red, sometimes orange, green
SHRUBBY SUCCULENT

This Madagascar native is shrubby, with many-branching stems covered in sharp half-inch-long thorns and bright green leaves. Bloom can be nearly year-round and features clusters of bright red or coral pink (occasionally yellow or orange) bracts at branch tips. Tolerates dry conditions as long as temperatures are not too low. 'Apricot' has flowers (bracts) that are apricot or coral blush.

Euphorbia × *Martinii* 'Ascot Rainbow'

Euphorbia milii

Euphorbia myrsinites
myrtle spurge

COLD TOLERANCE −20°F
INTEREST winter to early spring
SIZE ½–1 ft. tall, 1–2 ft. wide
COLOR chartreuse-yellow, blue green
PERENNIAL

Myrtle spurge is an almost-succulent with pointed, blue-green triangular leaves that look like scales arranged in a swirl along sprawling stems that emerge from the base. Early in the year, stem tips develop clusters of vivid chartreuse-yellow flower bracts that mark the first color of the garden's new year. Bloom is long, often from January to May. After bloom come reddish seedpods. As the seedpods dry out, they burst open with an audible "pop" as the seeds shoot out around the garden. This is the process of naturalization. After seeds are released and stems die back, cut stems off (wear protective clothing and glasses). New stems will appear shortly. Excellent edging plant.

Euphorbia rigida
gopher spurge

COLD TOLERANCE −20°F
INTEREST winter to early spring
SIZE 2–3 ft. tall and wide
COLOR bright green, blue green, chartreuse
PERENNIAL

Gopher spurge is similar to the compact *Euphorbia myrsinites*, but much larger and shrubbier, with stems that are more upright (hence the species name "rigida"), open, and airy. Leaves are longer, more lance-shaped, and spaced further apart along the stem. Best placed in the middle of a garden bed. Do not plant in New Mexico as this is considered a noxious weed in that state.

Euphorbia myrsinites

Euphorbia rigida

Euphorbia tirucalli 'Sticks on Fire'
red pencil tree, red pencil cactus

COLD TOLERANCE 30–35°F
INTEREST year-round
SIZE 8 ft. tall, 3–4 ft. wide
COLOR orange, yellow, red, green
SUCCULENT

The South African native pencil tree is grown for the color of its branches, not for flowers. This is a large upright succulent (though not a cactus) with narrow, pencil-thick branches. The standard pencil tree (which is blue-green) can grow quite large, but 'Sticks on Fire' is much smaller. On a cool-temperature, full-sun, low-water diet, the stems turn shades of bright yellow to orange-red.

Ferocactus
barrel cactus

Ferocactus is a genus of rounded, typically single, barrel-shaped cacti, all covered in tough, curved spines. Barrel cacti are native to the deserts of the southwest, from Arizona to Texas and south into Mexico. Their spines offer some protection from predators, while also shading the surface of the cacti. As these cacti mature, they form deep ribs beneath the spines. These ribs are like bellows that expand as the cacti absorb water after rains, then contract in dry times. Blooms occur in a ring toward the top of each barrel. Summertime flowers have papery petals in bright pink, yellow, red, or purple. When pollinated by cactus bees, flowers form yellow fruits. These cacti tend to live in areas where water accumulates from time to time.

Ferocactus pilosus
Mexican lime cactus

COLD TOLERANCE 25°F
INTEREST year-round
SIZE 8 ft. tall, 1–2 ft. wide
COLOR white, pale to blood red, orange, yellow, green
CACTUS

These Mexican natives have a dense covering of fine white spines, overlaid by a halo of larger and somewhat curved pale red to blood red spines. The white spines help shade the green surface of the deeply ribbed barrel. Orange, yellow, or red flowers encircle the very top of the barrel, followed by yellow fruits.

Euphorbia tirucalli 'Sticks on Fire'

Plant into lean, well-draining soils in the ground or in brightly glazed oxblood red, deep caramel, or cobalt blue pots. While full sun helps promote flowers in the hottest desert areas, these barrels appreciate some dappled midday shade at summer's peak.

Ferocactus wislizeni
fishhook barrel cactus

COLD TOLERANCE 5°F
INTEREST year-round
SIZE 6 ft. tall, 18–30 in. diameter
COLOR pale burgundy, green, brilliant gold or orange
CACTUS

These barrel cacti are really more cylindrical than round, and grow slowly to their mature height. Plants are covered in pale burgundy spines that are curved like hooks (in fact they are used as fish hooks by Native Americans). The body of the cactus is green with many ribs. In late summer, brilliant gold or orange flowers appear at the top of the barrel, followed by lemon yellow fruits. Grow in the ground or in deep ochre glazed pots in lean, well-draining soils with little to no irrigation. With tough love, these plants can live more than 100 years.

Ferocactus pilosus

Ferocactus wislizeni

Fouquieria splendens
ocotillo

COLD TOLERANCE 10°F
INTEREST spring and summer
SIZE 10–30 ft. tall, 15 ft. wide
COLOR fiery orange red, green
SHRUB

I think of this slow-growing Southwest and Mexican desert native as "flamethrower bush" because of its pointed clusters of fiery orange-red flowers that emerge from branch tips in spring and summer. This very sculptural woody plant is vase shaped with many upright branches that emerge from the base. Branches are covered in small clusters of round, leathery green leaves interspersed with long sharp thorns. Leaves drop in dry periods, and emerge following rainfall or garden irrigation. Ocotillos are best in desert or hot inland and valley gardens. Plant from bare root or container into full sun and well-draining soil. Do not irrigate. Attractive to hummingbirds.

Fouquieria splendens

Furcraea foetida 'Mediopicta'

COLD TOLERANCE 30°F
INTEREST year-round
SIZE 5–7 ft. tall, 6–8 ft. wide
COLOR green and cream
SUCCULENT

This large, upright agave relative has stiff, deep green- and butter yellow–striped, spear-shaped blades with no teeth or spines. Though large, it does not develop a trunk. After several years, it sends up a 25-foot-tall flower spike covered in cream-colored flowers that develop into tiny plants referred to as bulbils. When the stalk falls over, the bulbils scatter and root—or you can gather and root them into well-draining, sandy soils. The flowering process takes more than a year, after which the mother plant dies without making pups. Best in full sun in coastal gardens, or dappled shade inland. Water sparingly. When sited where the plant is backlit, the blades glow. (Note that there is a similar plant with the same name that stays smaller, 3–4 feet tall and 4–5 feet wide, with shorter, wavy blades, deep green with a lemon yellow center stripe.)

Gaillardia ×*grandiflora*
blanket flower

COLD TOLERANCE −30°F
INTEREST summer
SIZE 2 ft. mounds
COLOR red, orange, gold, yellow, gray-green
PERENNIAL

Blanket flower is named for its multicolored flowers that bloom in brilliant shades of yellow to orange to scarlet to rust. This mounding perennial has slightly fuzzy, deeply sculpted leaves, more gray-green than green. In summer's heat, mounds are covered in tall, slender flower stalks, each topped with a daisy-shaped flower with single or multiple rows of petals, some ruffled along the edges. 'Fanfare Blaze' is one of countless hybrids with different flower colors and patterns. Plant in full sun and lean soil, with little irrigation once established. Deadhead to extend bloom. These plants are extremely heat and drought tolerant. Allow to set seed to feed birds and so plants can reseed politely through the garden. It is also a favorite of butterflies.

Furcraea foetida 'Mediopicta'

Gaillardia ×*grandiflora*

Grevillea 'Long John'

COLD TOLERANCE 20–25°F
INTEREST spring and fall
SIZE 15 ft. tall and wide
COLOR green, watermelon, pale pink
SHRUB

Grevillea is a large group of Australian woody shrubs and trees in the family Proteaceae. Their showy flower clusters are made up of tiny, colorful, tube-shaped flowers, most an inch or two long. A long, curved or curled pistil, the female part of the flower, protrudes from each tube. In some varieties, the petals and pistil are a single color: bright red, maroon, raspberry coral, creamy white, buttery yellow, apricot, or bright gold. In others, the petals and pistil are different, making for multicolored flowers. All are hummingbird magnets. Like Ceanothus or Arctostaphylos, there are options from tree sized to groundcover scale. 'Long John' has large watermelon red and pale pink flowers and narrow, needle-like green leaves. Grow in full sun inland and along the coast in well-draining, lean soils with minimal irrigation once established, especially in summer (no irrigation if possible). Do not use fertilizer and especially do not use high phosphorus fertilizers, which will kill the plants. Excellent for cut flowers.

Hesperaloe parviflora
red yucca, yellow yucca

COLD TOLERANCE −20°F
INTEREST summer
SIZE 3 ft. tall and wide
COLOR blue-green or deep green, coral red or butter yellow
SUCCULENT

This hesperaloe is not an aloe at all, but rather a heat-loving succulent with 3-foot clumps of stiff, narrow, leathery, gray-green blades. In summer's heat, it sends up tall stalks of coral red or butter yellow tube-shaped flowers, much beloved by hummingbirds. Cut spent flowers once the bloom is over. Plant into well-draining soils in full sun, though in desert gardens it will also tolerate part shade. Water to establish, then infrequently if at all. Grow either color in a cobalt blue glazed container. Mass plantings are stunning. Keep an eye out for newer cultivars bred for more flowers, longer bloom, and more colors.

Grevillea 'Long John'

Hesperaloe parviflora

Hunnemannia fumariifolia
Mexican tulip poppy

COLD TOLERANCE 15°F
INTEREST spring to early summer, fall
SIZE 1½–2 ft. tall, 2 ft. wide and creeping
COLOR blue-green, buttery yellow, golden orange
PERENNIAL

Mexican tulip poppy looks like the sunnier, larger cousin to the bright orange California poppy. This poppy features fine, blue-green foliage on taller flower stalks, plus stems that creep horizontally to spread. Early spring flowers are the typical poppy cup shape with four silky petals. The lemon yellow petals serve as a bright backdrop for a fringe of golden orange stamens in the center of each flower. In cold winter areas, this Mexican desert native is grown as an annual. In warmer gardens, they live from year to year, reseeding gently through the garden, filling empty spaces with sunny spots of color. Plant from seed or container, in full sun, with little irrigation once established.

Hunnemannia fumariifolia

Iris

Bearded irises, along with California native irises (and their hybrids), are surprisingly drought tolerant. These perennials have slender upright blades, pointed at the tip, ranging from bright green to blue-green to variegated. The foliage is lovely and textural, but most gardeners grow iris for their fabulous springtime flowers. Iris flowers are typically large, some as big or bigger than a baseball and in color combinations that include yellow, purple, burgundy, apricot, mahogany, gold, white, almost blue, and brown, often with fancy markings. Iris flowers are complex, with two sets of petals: three curving upward (called "standards") and three downward ("falls").

Iris hybrids
bearded iris

COLD TOLERANCE 25°F or colder
INTEREST spring
SIZE 3 ft. tall, 8 in. wide
COLOR green, white, burgundy, purple, pink, almost black, blue, gold, soft yellow, and more, mostly in combinations
BULB/RHIZOME

Beaded irises make geometric fans of green blades with large flowers in spring—and some are rebloomers. The "beard" is a bit of fuzz at the base of the downward-curving petals. They prefer full sun and well-draining soil. Irrigate to establish and then on occasion during the active growing seasons; fertilize depending on your climate and location. These are more heat and drought tolerant than California native irises. They can go dormant in summer, depending on the climate, so don't be concerned if the leaves disappear once bloom is over. Allow

Iris hybrid

plants to go dry through summer. Underground, their fleshy rhizomes keep them going and even spread. As clumps expand, lift and divide every few years to keep plants flowering.

Iris Pacific Coast hybrid

COLD TOLERANCE 15–20°F
INTEREST spring
SIZE 10–12 in. tall and wide
COLOR white, burgundy, purple, yellow, almost brown, or combinations
BULB/RHIZOME

Cross two California native irises (*Iris douglasiana* and *I. innominata*) and you get Pacific Coast hybrid irises. These are more petite than the standard bearded iris—and lack the beard—but are equally beautiful. Flowers range from clear white ('Canyon Snow') to deep burgundy ('Dorthea's Ruby'), and purple ('San Ardo') to yellow ('Clincher'). Plant in cool sun along California's coast, or in afternoon shade in hotter inland gardens. Irrigate to establish and then on occasion during the active growing season. Divide clumps every few years.

Iris Pacific Coast hybrid

Kalanchoe

Though I've heard "kalanchoe" pronounced many ways, my botany professors said "kal-un-koh-ee," so I do, too. *Kalanchoe* is a very diverse group of flowering succulents, mostly shrubs and perennials. Garden kalanchoes have very interesting leaves in different shapes and colors. Some are smooth, some very fuzzy; some bright green, others silver, copper, pink, orange—if you lay them all out next to each other, you'd have a hard time understanding how they could be related, until you noticed the commonalities of their tube-shaped flowers. Some of those flowers are brightly colored and exuberant, others pale and demure. These succulents exhibit Crassulacean acid metabolism (CAM), a special adaptation to very arid conditions, where plant leaves open their stomates (leaf pores) in the cool of night to pull carbon dioxide from the atmosphere and release oxygen. It's a complicated process that helps limit water loss from the leaves to the atmosphere in hot, dry conditions. Kalanchoes are very easy to grow, whether in the ground or in containers. They are very decorative and add great color and texture to gardens.

Kalanchoe fedtschenkoi
lavender scallops

COLD TOLERANCE 25°F
INTEREST spring and summer
SIZE 15 in. tall and spreading
COLOR blue-green, coral pink
SUCCULENT

This is one of the prettiest shrubby kalanchoes, with small, upright blue-green blades, ruffled at the edges, that line relaxed stems. As the plant spreads, it fills empty spaces between nearby plants or softens the transition from wall to garden. In spring and summer, a patch becomes a mass of many flower stalks with hanging clusters of tubular coral pink flowers. Grow in a pot as a "spiller." 'Variegata' is a selection whose ruffled edges are cream with pink striations. Grow in light shade in desert gardens, full sun in coastal gardens, or full sun to light shade inland. Water occasionally once established. Also known as *Bryophyllum fedtschenkoi*.

Kalanchoe fedtschenkoi

Kalanchoe luciae
flapjack plant, paddle plant

COLD TOLERANCE 25°F
INTEREST year-round
SIZE 8–12 in. tall, 8 in. wide; forming colonies
COLOR ice green, yellow-green, orange, rose, rusty red
SUCCULENT

It's easy to see how flapjack plant got its name: its succulent blades look like a series of flat, rounded discs, or flapjacks, arranged on edge around a central stem. The discs are ice green about two-thirds of the way up, fading to yellow-green, orange, then rosy or rusty red. Somewhere in that sunset color combination is a band of pink, too. In spring, older plants develop a tall, ice green flower spike with yellowish flowers. After flowering, the mother plant dies back. Flowers are not the highlight of this plant; its vibrant blades, though, color up any garden bed or container. Grow in full sun along the coast, or light shade inland (the more light, the more intense their color). These are thirstier succulents that appreciate the occasional deep watering in the warm months, but they would rather be dry in cooler temperatures. Also known as *Kalanchoe thyrsiflora*.

Kalanchoe luciae

Kalanchoe pumila
flower dust plant

COLD TOLERANCE 25°F
INTEREST late winter to spring
SIZE 12–18 in. mounds
COLOR silver-gray, lavender-pink
SUCCULENT

Who wouldn't love a succulent with velvety purple and silver-gray leaves that sports clusters of lavender-pink flowers? This small shrubby kalanchoe is perfect for a cobalt blue glazed container, a rock garden, or cascading over a stone wall. Flowers appear in late winter or early spring and stick around for a while. Grow in light shade in desert gardens, full sun in coastal gardens, or full sun to light shade inland. Water occasionally once established.

Kalanchoe pumila

Lagerstroemia
crape myrtle

COLD TOLERANCE 0°F
INTEREST summer and fall
SIZE standard: 20 ft. tall and wide, dwarf: 6–15 ft. tall and wide
COLOR scarlet, orange, gold in fall; pink, purple, magenta, white in summer

TREE

Though crape myrtle is native to Asia, its heat tolerance makes it a go-to plant for hot, dry climates as well. This is a reliable summer bloomer for the West and Southwest. In summer, cranberry-colored buds open to magnificent sprays of crinkly petaled, raspberry red, deep lilac, pale pink, fiery red, or bright white flowers. Clusters of green fruits follow in early fall. The leathery green or burgundy leaves turn gold, orange, and copper before they drop in fall. Bare trees show off crape myrtle's fantastic mottled, peeling bark in shades of pink, pale green, red, and soft brown. Grow dwarf crape myrtle in containers or even as bonsai. Shop for trees in bloom to find your favorite color. For coastal gardens, look for mildew-resistant varieties. Plant in full sun in California gardens, or sun to light shade in Arizona. In New Mexico gardens, site carefully to protect from frost. In the first year, water deeply from spring through late fall. Starting the second spring, irrigate with the occasional deep watering. Always mulch.

Lagerstroemia 'Muskogee'

Lavandula
lavender

What's not to like about lavenders? These small evergreen shrubs or perennials have showy silvery gray to green, even variegated, fragrant foliage, beautiful purple, pale yellow, white, or raspberry flowers, and are extremely versatile garden plants. Mediterranean lavenders are best adapted to hot, dry conditions. Plant in fall through spring in areas where winter temperatures don't drop too far below freezing. In colder regions, wait until spring. Site in full sun in well-draining soil—poor soils are fine. Water regularly through the first season to establish; after that, allow Mediterranean lavenders to go dry between waterings. Overwatering can cause fatal rot. Fertilize in early spring with a balanced organic fertilizer, or don't fertilize at all. Most lavenders, with the exception of Spanish lavenders, require regular pruning. Few pests bother these plants, but they are very attractive to bees and other pollinators.

Lavandula allardii 'Meerlo'
Meerlo lavender

COLD TOLERANCE 23°F
INTEREST year-round
SIZE 2–3 ft. tall, 2½–3 ft. wide
COLOR green and butter yellow variegation
PERENNIAL

'Meerlo' is a recent introduction that is heat and drought tolerant. This fast-growing lavender doesn't seem to flower, but grow it for its incredibly fragrant and beautiful foliage. 'Meerlo' lights up a dark spot, contrasts solid-colored leaves, and is an all-around fabulous garden plant. Grow in full sun to part shade and well-drained soil. No need to fertilize.

Lavandula allardii 'Meerlo'

Lavandula ×*intermedia* 'Grosso'
purple lavandin

COLD TOLERANCE 15°F or cooler
INTEREST spring
SIZE 2–3 ft. tall, 3 ft. wide
COLOR lilac blue, gray-green
PERENNIAL

This hybrid Mediterranean lavender is one of the most fragrant and is used for perfume, cosmetics, and cooking. Narrow upright stems form a fountain-like mound. Deep lilac purple flowers top tall stalks and are excellent for cutting. Narrow leaves are gray-green. This lavender is short lived, from three to five years, but worth growing. Look for newer cultivars that are also hybrids of *Lavandula angustifolia* and *L. latifolia*.

Lavandula ×*intermedia* 'Grosso'

Lavandula stoechas
Spanish lavender

COLD TOLERANCE 10°F or lower
INTEREST spring to summer
SIZE 16 in. tall, 18 in. wide
COLOR deep purple, raspberry, white, pale yellow, bright green, silver-gray
PERENNIAL

Spanish lavenders are among the most drought resistant lavenders, and just as beautiful if not as aromatic as English and French lavenders. Blooms are distinct: dense, columnar flower clusters topped in colorful bracts are often referred to as "bunny ears" because they resemble just that. 'Wings of Night' blooms deep purple, but there are also raspberry, white, and pale yellow cultivars. Blooms spring through summer. Woody branches are lined with bright green or silver-gray needlelike foliage. Under the right conditions, Spanish lavenders produce a seedling or two.

Leucadendron
conebush

Conebushes are upright shrubs (along with a few trees) from South Africa characterized by leathery, oval or narrow and pointed leaves held upright and tight against the stems to the point of resembling scales. The leaves are the focus here, and they come in many shades of gray to green, red to yellow, and variegated. Conebushes make "flowers" from scales and small cones at branch tips (male and female on separate plants), but only in a few species or varieties are the cones notable. These plants thrive in Mediterranean-climate gardens in full sun and well-draining soils. They require no fertilizer and take little irrigation once established. Like many of their relatives, conebushes are susceptible to soil fungi mostly in summer, so avoid summer irrigation once plants are established. Grow as background, natural hedges, or specimens. Cut the branches for flower arrangements. Once they have some size on them; wherever you cut, many new branches will form.

Lavandula stoechas

Leucadendron discolor 'Pom Pom'
pom pom conebush

COLD TOLERANCE 20–25°F
INTEREST year-round
SIZE 6–7 ft. tall, 5–6 ft. wide
COLOR rusty orange, gold, golden yellow, blue-green
SHRUB

This is one of the few conebushes grown for its cone as well as its leaves. Male pom poms form large, bright rusty orange and gold cones that are surrounded by golden yellow bracts, all at the tips of branches covered in blue-green blades. Tolerates light shade.

Leucadendron 'Ebony'
ebony conebush

COLD TOLERANCE 20–25°F
INTEREST year-round
SIZE 3–4 ft. tall, 3–5 ft. wide
COLOR burgundy-black
SHRUB

This conebush has upright, deep burgundy (almost black) blades lining its branches. It is a beautiful garden accent when planted next to bright green succulent *Aloe ferox* or blue-bladed *Agave attenuata* 'Nova'.

Leucadendron discolor 'Pom Pom'

Leucadendron 'Ebony'

Leucadendron 'Jester'

jester conebush

COLD TOLERANCE 20–25°F
INTEREST year-round
SIZE 5–6 ft. tall, 3–5 ft. wide
COLOR red, green, yellow
SHRUB

'Jester' refers to the tricolor variegation on this conebush's leaves. The broad, upright leaves are striped green down the center and edged in red-blushed creamy yellow. Plant where 'Jester' will be backlit in the early morning or late afternoon. In that light, the colors glow. Also known as 'Safari Sunshine'.

Leucadendron salignum 'Winter Red'

winter red conebush

COLD TOLERANCE 20–25°F
INTEREST year-round
SIZE 4 ft. tall, 6 ft. wide
COLOR deep red, green
SHRUB

Selections of *Leucadendron salignum* tend to have deep red foliage (sometimes bright yellow), especially their new growth. 'Winter Red' is a smaller conebush, whose narrow blades are reddish in the cooler months, and more green in the heat. Grow in full sun or light shade in coastal and inland California gardens.

Leucadendron 'Jester'

Leucadendron salignum 'Winter Red'

Leucophyllum langmaniae 'Lynn's Legacy'
Lynn's legacy Texas sage

COLD TOLERANCE 10°F
INTEREST summer to early autumn, on and off
SIZE 4 ft. tall, 5 ft. wide
COLOR silver, purple
SHRUB

Texas sages are mounding evergreen shrubs with tiny gray-green to silver leaves. 'Lynn's Legacy' is a summertime bloomer that looks like a lavender-colored cloud when in flower. Native to the Southwest and Mexico. It performs and flowers better in desert and hot inland or valley climates than at the coast. Grow in full sun, well-draining soil, and with little irrigation. Also known as 'Lynn's Everblooming'.

Leucophyllum langmaniae 'Lynn's Legacy'

Leucospermum
pincushions

Pincushions, native to South Africa, feature exotic, fist-sized flowers in dazzling combinations of yellow, red, hot pink, fiery orange, watermelon, and apricot. These popular cut flowers grow on handsome evergreen shrubs used for specimens or informal hedges. Pincushion's "pins" are female and male flower parts that protrude from colorful tubes of curled petals. Blooms appear from February through early summer, depending on species or variety. Cut flowers fresh for arranging, or deadhead after bloom. Lovely, sculpted leaves are attractive year-round. Leaves are tough and leathery in bright green, deep green, or silver, and sometimes covered in a soft down that shades the leaves and helps conserve water in hot, sunny, arid environments. Pincushions prefer a sunny spot with lean, well-draining soil. Irrigate deeply, making sure to keep water off the leaves. After plants are established, water only infrequently during summer to avoid root rot. Mulch, and avoid fertilizer.

Leucospermum cordifolium 'Yellow Bird'
nodding pincushion

COLD TOLERANCE 25–30°F
INTEREST spring
SIZE 4–5 ft. tall, 6–8 ft. wide
COLOR golden yellow, gray-green
SHRUB

This pincushion has fist-sized, bright golden yellow flowers with green leaves that are scalloped along the outer edges.

Leucospermum cordifolium 'Yellow Bird'

Leucospermum reflexum
rocket pincushion

COLD TOLERANCE 25–30°F
INTEREST spring to summer
SIZE 8–12 ft. tall and wide
COLOR silvery gray, fiery orange or bright yellow
SHRUB

This most unusual pincushion has silvery gray foliage and flowers whose fiery orange "pins" point backward ("reflex") so they resemble shooting rockets. 'Luteum' is a yellow-flowered variety. Absolutely stunning.

Leucospermum 'Veldfire'
veldfire pincushion

COLD TOLERANCE 25–30°F
INTEREST spring
SIZE 6 ft. tall, 6–8 ft. wide
COLOR Orange, yellow, green
SHRUB

'Veldfire' flowers have orange-tipped petals with yellow "pins." Deep green leaves with gold tips line its branches.

Leucospermum reflexum

Leucospermum 'Veldfire'

Liatris punctata
gayfeather

COLD TOLERANCE 15°F or cooler
INTEREST late summer to fall
SIZE 1½–2 ft. tall, 2–3 ft. wide
COLOR lavender green
PERENNIAL

The tall, narrow, grasslike blades of gayfeather emerge in spring and soon develop tall stalks of fringelike lavender flowers. These colorful prairie and grassland perennials grow in lean, well-draining soils. After blooming, plants go dormant until the following spring when they sprout from a deep taproot. Gayfeather does best in full sun. It is tolerant of most soils, but prefers those that are well draining and coarse. Water deeply but only infrequently during the growing season. Attracts butterflies. Best in cold-winter climates.

Liatris punctata

×*Mangave* 'Macho Mocha'

COLD TOLERANCE 10°F
INTEREST year-round
SIZE 1–2 ft. tall, 4–6 ft. wide; flower stalk 8 ft.
COLOR purple and gray green
SUCCULENT

This *Agave* and *Manfreda* hybrid has wide rosettes of thick, fleshy, pliable gray-green leaves covered in purplish spots. A thick, tall flower stalk rises from the center. Its many whitish flowers attract hummingbirds. After flowering, rosettes make pups. Plant in full sun to light shade and water infrequently. Watch for snails. Perfect for growing in masses in garden beds or as a single specimen in a container.

×*Mangave* 'Macho Mocha'

Mascagnia macroptera
Mexican orchid vine

COLD TOLERANCE 20–25°F
INTEREST spring and summer
SIZE 10–20 ft. long
COLOR yellow
VINE

Mexican orchid vine features fleshy, bright green leaves and sunny yellow, orchid-shaped flowers. When petals fall, they leave behind chartreuse green seedpods that look like butterfly wings. This climbing, evergreen vine is very drought tolerant in California, but thirstier in Arizona. Grow over a wall, on a strong trellis, or covering a pergola. Plant in well-drained soil and full sun along coast and inland, or some shade in desert regions.

Mascagnia macroptera

Melianthus major
honeybush

COLD TOLERANCE 15–20°F
INTEREST spring and year-round
SIZE 6 ft. tall, 8 ft. wide
COLOR burgundy, gray-green, or rose-blushed purple
SHRUB

Honeybush is a highly sculptural shrub with many tall corky stems, each supporting wide, frosty blue leaf fronds. The frost is from a waxy coating that protects

Melianthus major

the leaves from water loss. In spring, each stem produces a comb of burgundy flowers filled with sweet nectar that attracts hummingbirds. Don't worry if some nectar spills out onto leaves and attracts black sooty mold. Plant honeybush away from walkways or places where people can brush up against the leaves to release their musty, rancid peanut butter fragrance—not bad enough to keep it out of the garden, of course. Cut stems to the ground as flowers fade. Soon, more stems emerge. 'Purple Haze' is a 3-foot-tall dwarf cultivar with rose-blushed purple fronds. Offers fabulous color in the garden.

Mimulus
monkey flower

Monkey flowers are small flowering shrubs or perennials whose characteristic flowers are long trumpets, broadly flared at the end and said to resemble a smiling monkey's face (I don't see it, but that's okay). Most have green leaves and flower in shades of white, yellow, gold, pink, red, or rose. Hybrid monkey flowers sold as annuals or short-lived perennials in nurseries are bred from moisture-loving parents. Those listed here are selections or hybrids of California native monkey flowers chosen for their drought tolerance in arid gardens. Some monkey flowers go dormant in summer as a strategy to survive the hottest and driest time of year. At that point you can cut them back by a third or so. If watered occasionally through summer, some stay evergreen.

Mimulus bifidus 'Esselen'
Esselen monkey flower

COLD TOLERANCE 15–20°F
INTEREST spring
SIZE 2–3 ft. tall and wide
COLOR orange, green
SHRUB

Mimulus bifidus 'Esselen' is native to California's Big Sur coastline and features very large bright orange flowers over a long bloom period.

Mimulus bifidus 'Esselen'

Mimulus 'Changeling'
changeling monkey flower

COLD TOLERANCE 15–20°F
INTEREST spring
SIZE 3 ft. tall and wide
COLOR rose, peach, bright green
SHRUB

This monkey flower features a bloom with an unusual mutabilis-pattern (changeable or varying) flower; rose in the center, fading to a frilly peach edge.

Mimulus 'Jelly Bean' hybrids
jelly bean monkey flowers

COLD TOLERANCE 20–25°F
INTEREST spring or year-round
SIZE 2–3 ft. tall and wide
COLOR purple, yellow, dark pink, bright green
SHRUB

The Jelly Bean hybrids are named according to their vibrant colors: 'Jelly Bean Orange', 'Jelly Bean Purple', 'Jelly Bean Lemon', 'Jelly Bean Dark Pink', and so on. Their glossy green leaves contrast beautifully with their plentiful flowers that are present almost year-round in coastal California gardens. Grow in full sun or light shade.

Mimulus 'Changeling'

Mimulus 'Jelly Bean Orange'

Muhlenbergia capillaris
pink muhly grass

COLD TOLERANCE 15°F
INTEREST fall
SIZE 2–4 ft. mounds
COLOR soft green, purple
GRASS

Few grasses are as colorful as pink muhly grass. Year-round, it features mounds of soft green and blue blades. In late summer or early fall, it is covered by a gorgeous purple-haze bloom. 'Regal Mist' is a particularly floriferous selection. Plant in full sun, part shade in deserts, and in almost any soil. If desired, cut back to 6 inches after bloom fades.

Muhlenbergia capillaris

Neomarica caerulea

walking iris

COLD TOLERANCE 25°F
INTEREST summer
SIZE 4–5 ft. tall, 3–4 ft. wide
COLOR green, cerulean blue–violet, gold, brown, white
BULB/RHIZOME

This evergreen member of the iris family shows off with tall fans of upright green leaves, topped in summer by taller stalks of large irislike flowers. Each flower features three broad cerulean blue–violet petals ("falls") with gold and brown ripples toward the flower's center. Three more upright petals echo the ripples in brown and violet on a white background. Flower stalks are lined in multiple buds, which open in sequence for just one day each. This is a show stopper when in flower, and offers excellent garden texture year-round. Grow in dry conditions in coastal and inland gardens, in cooler sun or light shade and well-draining soil. Good under the dappled light of a tree canopy. Rather than make seed, flowers make new plantlets that can be separated from the mother stalk and rooted into a pot once they begin to develop some roots.

Opuntia violacea var. macrocentra

tuxedo spine prickly pear, black spine prickly pear

COLD TOLERANCE 10°F or lower
INTEREST year-round, especially spring and summer
SIZE 3 ft. tall, 4 ft. wide
COLOR dusky purple, blue-green, black (spines), yellow, blood red
CACTUS

The very vertical and round, flat pads of this prickly pear are blue-gray some of the year, turning purple in the coldest and driest months. Year-round, the pads are covered in clusters of long black spines with white tips. Flowers are brilliant yellow tissue-paper petals with a blood red center. Flowers open during the day and close at night. This is a smaller prickly pear that thrives in lean, well-drained soils in full sun in New Mexico, Arizona, and the deserts of California. Also known as *Opuntia macrocentra*.

Neomarica caerulea

Opuntia violacea var. macrocentra

Parkinsonia 'Desert Museum'
desert museum palo verde

COLD TOLERANCE 15°F
INTEREST spring to summer
SIZE 25–30 ft. tall and wide
COLOR bright green, bright yellow
TREE

'Desert Museum' (a reference to the Sonora Desert Museum in Tucson) is the best and most beautiful of the palo verdes, the "green stick" trees from the deserts of the Southwest. Compared to other palo verdes, this hybrid is thornless, upright, graceful, long blooming, has larger flowers, and is more aesthetically pleasing in all ways. Quarter-sized bright yellow flowers put on their main show in spring, then do a repeat performance periodically through summer. Leaves are tiny and round. Trunks and branches are bright green and very sculptural. Plant in full sun and lean soil, with little to no irrigation. Grows fastest in the heat of the desert. In gardens within two miles of the coast it mildews and fails in the moist air. Heavily visited by bees.

Parkinsonia 'Desert Museum'

Plant Directory

Pedilanthus macrocarpus
slipper plant

COLD TOLERANCE 30°F
INTEREST spring, fall, year-round
SIZE 3–4 ft. tall, 2 ft. wide
COLOR coral or coral pink flowers
SUCCULENT

Slipper plant is a euphorbia relative with lime green succulent stems that are pinky finger thick. The occasional green leaf is small in sun-grown plants, and surprisingly large in plants grown in light shade. In spring and fall, primarily, stem tips sprout coral orange or coral pink bracts that look like open bird beaks (rather than slippers). Slipper plant is excellent as an upright spot of bright green and coral, used as a sculptural accent, or in a large container (oxblood red or teal glazed pots offer striking color contrasts). Plant in full sun or light shade. In sun, stems grow upright and compact; in shade, they grow wavy and longer. Offer little if any irrigation, and no pruning. Hummingbirds are frequent visitors. Look also for its larger and showier cousin, *Pedilanthus bracteatus*.

Pedilanthus macrocarpus

Pelargonium sidoides
South African geranium

COLD TOLERANCE evergreen above 20 or 25°F; goes dormant in cooler temperatures
INTEREST spring, summer, year-round
SIZE 1 ft. tall and wide
COLOR blue-green, burgundy
PERENNIAL

From South Africa comes this small, tough geranium that works beautifully as an edging plant in garden beds. Each mound is a mass of heart-shaped, nickel-sized, blue-green leaves. From the center rise many 6- to 10-inch-tall flower stalks topped in slim-petaled, red-burgundy, velvet flowers. Bloom peaks in spring and early summer, continuing on and off through the year in coastal gardens. Does well in full sun, and tolerates most soils; no fertilizer. Over time, stems tend to grow leggy, which suits container plantings when stems cascade over the side of a pot (try a teal blue glazed pot), but for in-ground plants, wait for a break in the bloom, then cut stems back to just a few inches long. Water well. Plants will soon regrow. Cuttings are easy to root.

Pelargonium sidoides

Penstemon

All but one species of *Penstemon* are native to North America. They feature colorful, tube-shaped flowers that attract hummingbirds, bees, and butterflies. Most are soft-stemmed perennials, though some develop some wood. Penstemons that are evergreen in milder gardens can die back in winter in colder regions; where they don't die back, they benefit from a good haircut after they finish flowering. Penstemons are not long lived, especially when overwatered, which is a possibility for the drought-adapted species included here. Skip the many thirsty hybrids sold by nurseries; try these instead and water sparingly. Space plants so they have plenty of room to spread—some even reseed.

Penstemon ambiguus
bush penstemon, sand penstemon

COLD TOLERANCE −20°F
INTEREST spring to fall
SIZE 3 ft. tall and wide
COLOR pale pink or white, green
PERENNIAL

From spring to fall, this fast-growing, shrubby Southwest native is covered in masses of pale pink to nearly white tube-shaped flowers. Tiny leaves are green and narrow. In the coldest areas, plants die back to the ground in winter. In milder-winter areas, they are evergreen. Grow in the hottest gardens, in full sun and alkaline soil with excellent drainage. Be careful not to overwater. Does not do well in containers.

Penstemon ambiguus

Penstemon eatonii
firecracker penstemon

COLD TOLERANCE −20°F
INTEREST late winter, early spring
SIZE 1–4 ft. tall, 1–2 ft. wide
COLOR deep green, firecracker red
PERENNIAL

Firecracker penstemon, another Southwest native, offers the best late-winter and early-spring color, with its firecracker red flowers on upright flowering stems and deep green leaves. This plant thrives in full sun and some shade in well-draining soils. Water only occasionally but deeply through the year, taking care not to overwater.

Penstemon heterophyllus
foothill penstemon

COLD TOLERANCE −10°F
INTEREST spring to summer
SIZE 1–3 ft. tall, 2–3 ft. wide
COLOR green, blue, rose purple or lavender
PERENNIAL

Penstemon heterophyllus is native to the chaparrals of California's foothills, where its showy blue and rose-purple blooms cover 1- to 2-foot-tall plants from spring through summer. 'Margarita BOP' is a somewhat smaller selection (2 feet by 2 feet), with the showiest and most colorful blue and lavender flowers. Alas, beauty has a price, and 'Margarita BOP' tends to be shorter lived in gardens and less drought tolerant than the standard foothill penstemon. Both do best in full sun, but tolerate light shade.

Penstemon eatonii

Penstemon heterophyllus 'Margarita BOP'

Penstemon triflorus
hill country penstemon, Heller's penstemon

COLD TOLERANCE 10°F or lower
INTEREST spring
SIZE 1½–2 ft. tall, 1–2 ft. wide
COLOR clear pink with deep pink and white, green, bronze
PERENNIAL

This is a Texas native penstemon with upright stems and deep green leaves, sometimes has a bronzy tone. All through spring, tall linear flower stalks rise past the foliage to show off their intense 2-inch pink flowers. Flowers feature white and deep pink markings, and are organized in clusters of twos or threes. Grow in lean soils, whether sand, loam, or clay (taking care not to overwater), and full sun or light shade.

Pilosocereus pachycladus
blue columnar cactus

COLD TOLERANCE 30°F
INTEREST year-round
SIZE 10–30 ft. tall, 4 in. diameter
COLOR sky blue, white, deep purple
CACTUS

This tall columnar cactus is native to Brazil's dry forest, and one of nature's few true blue plants. Its sky blue stems are accented in white spines and "fur" along its many ridges. Crepe paper, cream white flowers emerge from large, midnight blue or purple buds. This is a spectacular plant all in all for its color combination as well as its architecture. It is frost sensitive and prefers full sun with regular water in Arizona summers, and occasional irrigation in warmer months in California.

Penstemon triflorus

Pilosocereus pachycladus

Punica granatum

pomegranate (edible and ornamental, dwarf and standard)

COLD TOLERANCE 15°F
INTEREST spring and fall
SIZE standard: 15 ft. tall and wide; dwarf: up to 3 ft. tall and wide
COLOR vermillion orange, deep red, gold, green
SHRUB

Brilliant vermillion red spring flowers, large deep red (or soft pink or purple or yellow depending on variety) fruits in late summer or fall, green leaves that turn gold in fall—pomegranate is a near-perfect drought-tolerant shrub or small tree. Full-sized pomegranate shrubs produce edible fruits, from tennis ball to softball size. Dwarf varieties are bred for size, not fruit, so while they might produce fruits, don't expect them to be tasty. Pomegranates are excellent for growing in containers (teal blue, cobalt, or green glazed). Prune when dormant to remove dead branches to keep plants from looking ragged, or if you prefer a tree shape. It can take five to seven years after planting for fruiting to reach full production. Patience, please.

Punica granatum

Rosa Flower Carpet

Flower Carpet roses

COLD TOLERANCE −10°F
INTEREST spring and fall
SIZE 3 ft. tall and wide
COLOR green with peach or red and yellow
SHRUB

If you struggle with the idea of growing roses in a dry garden, here are some to try. Flower Carpet roses are among the most drought-tolerant roses, especially in coastal Mediterranean climates. These roses have shiny green leaves and bloom from spring through fall in California gardens; in early spring and again in fall in desert gardens. Their 2½-inch-diameter flowers grow in clusters. 'Amber' blooms peach or apricot and is lightly fragrant; 'Red' has brilliant yellow stamens; and there are other colors as well. In desert gardens, plant in light shade, full sun elsewhere. These roses prefer well-amended soils, but will tolerate lean soils. Water to establish during the first year, then only occasionally in coastal and valley gardens, more often in desert gardens. Fertilize twice yearly and cut plants back by two-thirds in late winter or early spring.

Rosa Flower Carpet 'Red'

Rosmarinus officinalis
rosemary

COLD TOLERANCE 10°F or cooler depending on variety
INTEREST year-round
SIZE upright: 4–6 ft. tall, 4–5 ft. wide; prostrate: 1–3 ft. tall, 6–8 ft. wide
COLOR deep green or green or cream variegated, blue, pale blue, white, or pink
SHRUB

Rosemary is more than an herb for cooking, it is a stalwart of Mediterranean gardens. This woody evergreen grows upright as a shrub or prostrate or cascading over a stone wall as a groundcover. Its narrow, needlelike deep green (occasionally variegated) leaves contain the oils that give rosemary its distinctive resinous fragrance and flavor—small, leathery leaves high in oils are characteristic dry-climate adaptations. Rosemary's small flowers are plentiful, in shades of blue, sometimes white or soft pink. Rosemary blooms from spring through fall in California and Arizona gardens, early spring and again in fall in New Mexico gardens. Prostrate rosemary plants make great groundcovers. Plant in full sun in California gardens, sun or part shade in the hottest deserts, and in well-draining soils. Water to establish, then deeply but infrequently. No need to fertilize. 'Tuscan Blue' and 'Arp' are cold-tolerant cultivars.

Rosmarinus officinalis

Ruellia peninsularis
desert ruellia

COLD TOLERANCE 25°F
INTEREST summer
SIZE 4–5 ft. tall and wide
COLOR purple, green
PERENNIAL

This rounded, evergreen desert native has glossy, triangular deep green leaves and is covered with blue or purple flowers that resemble small petunias. Bloom peaks in spring, then continues on and off into fall. Plant in full sun and well-draining soil. Water to establish and infrequently after that. If branches get frosted, wait until spring to cut back damage. Good choice for containers. Attracts butterflies, hummingbirds, and other birds.

Salvia
sage

Salvias are evergreen perennials and some shrubs found on nearly every continent and climate of the world. They all share characteristic square stems and two-lipped flowers arranged in spires. Many have aromatic leaves. When salvias grow too leggy or too woody, cut them back just to where new sprouts are forming along the branches. Mexican, chaparral, and desert sages are some of the best arid-climate garden plants. Plants in this genus are magnets for hummingbirds and butterflies. Many people promote *Salvia greggii* (autumn sage) for arid gardens, but, while it does survive in these climates, I find that it's too short lived and its branches too brittle. The sages included here are longer lived, tougher, and more reliable in a garden situation.

Ruellia peninsularis

Salvia 'Bee's Bliss'
bee's bliss sage

COLD TOLERANCE 15°F
INTEREST spring
SIZE 1–2 ft. tall, 5–8 ft. wide
COLOR gray green, lavender
SHRUB

If you are looking for a good slope plant, try *Salvia* 'Bee's Bliss'. This cross between *S. clevelandii* and *S. sonomensis*, both California natives, forms a low, broad blanket of aromatic, gray-green leaves that are beautiful year-round. In spring, the blanket is covered in lavender-colored flowers. Grow in full sun and well-draining soil, and water sparingly after establishment in California gardens.

Salvia chamaedryoides
germander sage

COLD TOLERANCE 10°F
INTEREST year-round
SIZE 18 in. tall, 2–3 ft. wide
COLOR silver green, deep green, clear blue
PERENNIAL

This Mexican native makes an excellent ground-cover that spreads by underground roots to form patches. Deep green or silvery leaves are tiny and rounded, with light fuzz. This sage is covered in clear blue flowers most of the year in California gardens, and spring through fall in desert gardens. Plant in full sun and lean soils, and water to establish. In my coastal garden, dry-grown plants and irrigated plants grow and bloom equally well. Combine with *Calylophus drummondianus* (sundrop) and *Kalanchoe luciae* (flapjack plant) for a beautiful textured, yellow-orange-blue-gray edging.

Salvia 'Bee's Bliss'

Salvia chamaedryoides

Salvia clevelandii
Cleveland sage, chaparral sage

COLD TOLERANCE 10°F
INTEREST spring
SIZE 4–5 ft. tall, 4 ft. wide
COLOR gray green, lavender, rich violet, maroon
SHRUB

Cleveland sage has highly aromatic soft green leaves that perfume the chaparral. This sage develops woody stems with lavender flowers. 'Winnifred Gilman', which is a more compact selection, flowers rich violet on maroon-colored stems. From spring through summer, tiny, two-lipped flowers appear in whorls, layered around a central stem. Grow in full sun. Provide minimal irrigation in California gardens (it is a California native), monthly winter watering in the desert, and just occasionally in summer. Well-draining soils are best. Hummingbirds visit the flowers, as do bees and other wildlife. Leave flower heads to make seeds for birds, then deadhead and cut branches back by up to a third, taking care not to cut into the wood.

Salvia dorrii
desert sage

COLD TOLERANCE –10°F or colder
INTEREST spring to summer
SIZE 1–3 ft. tall and wide
COLOR gray green, blue, raspberry
SHRUB

Desert sage is native to all of the western states, where it grows in open areas or foothills, in sandy and rocky soils. This smaller sage has upright stems covered in small, aromatic, oval gray-green leaves. The leaves form a backdrop for stalks of blue flowers with raspberry-colored flower bracts. Blooms from spring into summer. Plant in full sun in well-draining soil, and water to establish, then hold back.

Salvia clevelandii

Salvia dorrii

Salvia leucantha
Mexican bush sage

COLD TOLERANCE 20°F
INTEREST spring and fall
SIZE 3–4 ft. tall and wide
COLOR deep purple, white, deep green
PERENNIAL

This classic southwestern perennial sage has long wands of purple and white flowers that are recognizable in gardens from coast to desert. Flowers appear in spring and fall in California gardens, and summer through fall in New Mexico and Arizona. Mexican bush sage has upright stems covered in white fuzz that contrast the rough-surface deep green leaves. In colder climates, stems freeze back in winter. In milder climates, spent flower stalks look ratty, so most gardeners prune them back to the ground. New stems soon resprout. Grow in full sun, with little irrigation in all but the hottest and driest gardens. 'Santa Barbara' is a selection that grows about half the size of the species. 'Midnight' has all-purple flowers.

Salvia leucantha

Salvia microphylla 'Hot Lips'
hot lips little leaf sage

COLD TOLERANCE 0°F
INTEREST spring to fall
SIZE standard: 4–5 ft. tall and wide; dwarf: 18 in. tall and wide
COLOR bright green, red, white
PERENNIAL

Fast-growing *Salvia microphylla* has small, round green leaves and bright-colored showy flowers spring through fall. 'Hot Lips' blooms prolifically with an unusual combination of red and white bicolor flowers. A dwarf version called 'Little Kiss' has the same blooms on far smaller plants. Other selections bloom shades of pink to red. Grow in full sun or part shade. Plants flower best in full sun with well-drained soil. Irrigate well to establish, then cut back.

Salvia microphylla 'Hot Lips'

Salvia 'Pozo Blue'
pozo blue sage

COLD TOLERANCE 5°F
INTEREST spring
SIZE 3–5 ft. tall and wide
COLOR gray green, dusky purple blue
PERENNIAL

'Pozo Blue' has fragrant gray-green leaves and spires of dusky purple-blue flowers. This hybrid of two California native sages is extremely drought tolerant and does well in many soil types, from sand to clay. Grow in full sun and with little water once established.

Salvia 'Pozo Blue'

Salvia spathacea
hummingbird sage

COLD TOLERANCE –10°F
INTEREST spring and fall
SIZE 1–3 ft. tall, spreading
COLOR bright green, jewel magenta
PERENNIAL

Few plants flower in dry shade as well as hummingbird sage, which grows naturally under the canopy of oak trees. This sage has underground stems that spread to make patches, several feet across. Leaves are broad and bright green and release their fruity fragrance when bruised. In spring and fall, flower stalks rise above the foliage, and are covered in tiny, jewel magenta flowers. Grow in full sun along the coast, in shade everywhere else. Irrigate to establish, and occasionally thereafter. Cut back if plants grow so leggy that the centers die out. Hummingbirds fight over this plant.

Sedum nussbaumerianum
coppertone stonecrop

COLD TOLERANCE 28°F
INTEREST year-round
SIZE less than 1 ft. tall, 2–3 ft. wide
COLOR bronze, yellow, pink, apricot, orange, gold-orange
SUCCULENT

There aren't many other garden plants with golden orange foliage. This succulent's tiny chunky blades grow along creeping stems to form a mat. Grow this for the foliage, not for the tiny white flowers that appear from time to time. Excellent as a bedding plant to complement plants with gray-green or blue-green leaves. Great in mixed containers. Grow in full sun for best color, away from foot traffic (leaves fall off easily, but they root to make new plants); water sparingly. Sometimes confused with *Sedum adolphii*.

Salvia spathacea

Sedum nussbaumerianum

Senecio mandraliscae
blue chalk fingers

COLD TOLERANCE 15°F
INTEREST year-round
SIZE 1 ft. tall, 3–4 ft. wide
COLOR blue
SUCCULENT

Senecio mandraliscae has succulent stems that spread quickly and horizontally to form broad swaths. The stems and their upright succulent blades are blue, covered with a fine layer of white wax. This plant is great for creating a low, blue edging or setting the stage for a blue garden. *Senecio talinoides* has taller upright stems, more slender blades, and looks more greenish than blue. *Senecio serpens* looks very similar to *S. mandraliscae* but is more diminutive and not as vigorous. Whichever you grow, the flowers are not attractive so cut them off as flower stalks start to develop. Plant in full sun or part shade. Best in well-draining soils, but tolerant of clay.

Solidago spathulata var. nana
dwarf goldenrod

COLD TOLERANCE 0°F or colder
INTEREST summer and fall
SIZE 1 ft. tall and wide
COLOR yellow, green
PERENNIAL

This Pacific Northwest perennial is native to deserts where it develops tongue-shaped deep green leaves that form a low rosette and tall stalks of bright yellow fringy flowers. Grow where rainwater collects, in full to part sun. Also known as *Solidago simplex* var. *nana*.

Senecio mandraliscae

Solidago spathulata var. *nana*

Sophora secundiflora
Texas mountain laurel, mescal bean

COLD TOLERANCE 10°F or colder
INTEREST spring and year-round
SIZE 15–25 ft. tall, 15 ft. wide
COLOR green or silver-gray, lavender or deep purple
SHRUB OR TREE

This lovely New Mexico and Texas native is a small, slow-growing evergreen tree or large shrub, suited to hot gardens from Texas to Southern California. Its large leaves are made of many small and round leathery leaflets, each glossy green or silver. In early spring, the branches are covered in large hanging clusters of lavender, wisterialike flowers. Their grape bubblegum fragrance is unmistakable. Flowers are followed by flocked gray seedpods that hold bright red seeds (bright red can be a warning in the plant world and in this case it means the seeds are toxic). 'Silver Peso' and 'Silver Sierra' are both selections with silvery gray leaves and intense purple flowers. Plant against a deep-toned wall as a hedge or specimen plant in the garden. Grow in full sun to part sun in well-draining, alkaline soils. Water to establish, then deeply but sparingly after that, mostly in the warm months of the year.

Sophora secundiflora

Sphaeralcea ambigua
desert mallow

COLD TOLERANCE −10°F
INTEREST spring to summer
SIZE 2–4 ft. tall and wide
COLOR silver-green, apricot, white, pink, pale purple, or watermelon red
SHRUB

This small, open shrub is a true western native with small, silvery green leaves covered in fuzz. In early spring, cuplike flowers cover the branches, most often in shades of apricot, but sometimes white, pink, or pale purple. 'Louis Hamilton' is a selection with watermelon red–colored flowers. Some years, plants bloom again after summer. In early fall, thin out dead branches, then cut the plant back to 6 inches; a flush of new growth will follow. As old plants die out, new ones sprout from seed to replace them, ensuring their presence year after year. Plant in full sun in well-draining soils. Water to establish; after establishment, plants need little if any irrigation in California gardens, periodic irrigation in Arizona and New Mexico gardens.

Sphaeralcea ambigua

Tecoma stans
yellow bells

COLD TOLERANCE 5°F
INTEREST spring to fall
SIZE 5–12 ft. tall, 6–12 ft. wide
COLOR bright green, yellow, orange, red, and combinations

SHRUB

Gardeners are surprised to learn that this showy deciduous shrub is native to deserts from the Southwest into South America. It is a fast grower with glossy green leaves covering lose branches. Yellow bells grow taller in California and Arizona mild-winter gardens. In New Mexico grow as a perennial in-ground or in a container that can be moved to a protected location in winter. From spring through fall, the clusters of large, trumpet-shaped flowers draw hummingbirds and butterflies. The species has bright yellow blooms, but plant explorers and breeders have been working hard to find new color selections. Look for the smaller-stature *Tecoma stans* var. *angustata*, which has bright yellow flowers, along with hybrids such as 'Bells of Fire', whose fiery red blooms turn deep orange in summer's heat, or 'Crimson Flare', also with red flowers, though these turn pink in the heat. 'Sunrise' is a personal favorite; the outer surfaces of its bells are bright orange fading to golden yellow. The trumpet's open flares are yellow too, blushed orange with orange lines into the throat. The colors are luscious. Plant in full sun and nearly any soil. Water to establish, then only periodically in the growing season. After the bloom is over, prune to remove spent flower stalks and shape plants so they don't grow too rangy. *Tecoma capensis* (formerly *Tecomaria capensis*) is *not* recommended for California gardens, where it is an extremely aggressive spreader.

Tecoma stans

Verbena

Verbenas are perennials valued for their green leaves and colorful clusters of tiny, bright-colored flowers. There are many thirsty hybrid verbenas sold by nurseries. Those included here are verbenas that are drought tolerant, and all colorful. Plants in this genus are frequented by bees and butterflies.

Verbena bonariensis

tall verbena

COLD TOLERANCE 10–15°F
INTEREST spring to fall
SIZE 3–6 ft. tall, 2 ft. wide
COLOR deep green, violet purple
Perennial

Tall verbena is an airy perennial that grows many tall green wands, each topped in dome-shaped, half-dollar-sized clusters of tiny violet purple flowers from spring through fall. Leaves are deep green, rough, and scratchy, especially as they dry at the end of the season, so site this plant away from walkways. Despite its perennial status, it seldom survives winter, though it does reseed. Plant amid grasses in a meadow, as vertical accents in a perennial bed, or as a color punctuation mark against silver-colored agaves. Might be weedy in New Mexico.

Verbena bonariensis

Verbena lilacina 'De la Mina'
lilac verbena

COLD TOLERANCE 25°F
INTEREST spring and summer
SIZE 1–2 ft. tall, 2–3 ft. wide
COLOR pale lilac, deep green
PERENNIAL

This beautiful verbena is native to Cedros Island off the west coast of Baja California. In bloom, it looks like a small purple cloud. Nickel-sized, domed flower clusters are made up of many small, fragrant violet–colored flowers. Bloom peaks in spring but continues on and off through the year in California gardens. Full sun in coastal gardens, light sun or part sun in hot inland gardens. Well-draining soils are best, but will tolerate heavy soils. Give it a deep watering monthly from spring through fall.

Verbena lilacina 'De la Mina'

Verbena rigida
sandpaper verbena

COLD TOLERANCE 5°F
INTEREST spring and summer
SIZE 1 ft. tall, 3–4 ft. wide
COLOR bright green, deep purple
PERENNIAL

This spring- through fall-blooming perennial verbena grows well in desert gardens. The name "sandpaper verbena" comes from its rough, scratchy green leaves. Below ground, creeping underground rhizomes can "move" the main plant some feet from where it was first planted. In spring, clusters of tiny deep purple flowers appear, creating a beautiful color contrast. Grow as a groundcover in full sun, and water occasionally once established. If plants start to look ratty, cut them back to the ground; they will resprout.

Verbena rigida

Vitis 'Roger's Red'
Roger's California grape

COLD TOLERANCE 0°F
INTEREST fall
SIZE 40 ft. long
COLOR crimson red
VINE

Like all grapevines, 'Roger's Red' grows vigorously and is deciduous. Before it drops its foliage in fall, however, the leaves turn from gray-green to a fabulous glowing crimson red. Though grown as an ornamental, the vine does ripen grapes in summer. Their large seeds and bitter skins make them best suited to birds and other wildlife. Grow in full sun, part sun, or light shade. Provide strong support for the vine's heavy stems, though dropping fruits and leaves make this a poor choice for pergolas and patio covers. Instead, grow over a chain-link fence or trellis. Established plants need very little if any irrigation. Train and prune as desired.

Yucca 'Bright Star'

COLD TOLERANCE 0°F
INTEREST year-round
SIZE 2 ft. tall, 3 ft. wide
COLOR yellow, green
SUCCULENT

'Bright Star' is a small, trunkless (or nearly so) yucca whose blades are butter yellow with pastel green striations down the center. In cooler temperatures or in drought, leaves take on a pinkish blush. They are slightly arched and tipped with a sharp spine. While 'Bright Star' grows slowly, its bright coloration lights up a dry garden. In late summer, it forms tall stalks with pink buds that open to white flowers, but they are very much secondary to the foliage. Plant in masses or grow as a single specimen in cobalt blue, oxblood red, or caramel glazed containers. Grow in full sun, in sandy, well-draining soil. Irrigate occasionally, but deeply.

Vitis 'Roger's Red'

Yucca 'Bright Star'

Zinnia grandiflora

desert zinnia, prairie zinnia

COLD TOLERANCE −30°F
INTEREST summer
SIZE 1 ft. tall and wide
COLOR green, golden yellow
ANNUAL

When it comes to desert fillers and groundcovers, *Zinnia grandiflora* is one of the best. This perennial zinnia blooms from spring into fall with bright, quarter-sized golden yellow flowers. Many stems emerge from a common base, each densely covered with narrow, bright green leaves. Below ground, the stems spread by rhizomes, ultimately reaching up to 10 feet across though less than a foot tall. Plant in full to part sun. Water occasionally during bloom, but take care not to overwater, especially for plants growing in clay soils. Though used widely in desert gardens, this zinnia is not widely planted in California gardens but certainly is worth trying.

Zinnia grandiflora

Bromeliads

Bromeliads are tropical and subtropical plants, some terrestrial (grow in the ground) while others are epiphytes that grow on other plants. All have stiff, leathery leaves that form rosettes or vases, that can be tiny (air plants are bromeliads) or up to 4 feet tall. Bromeliads are valued for their colored and patterned leaves that can be red, silver, burgundy, orange, pink, gray, blue-green, yellow, and dusky green, and striped, spotted, or variegated. Some have smooth edges while others are toothed or even edged in wicked spines. With that much variation, flowers play second fiddle, but what beautiful second fiddles they are. Their colors are exotic: combinations of glowing pink, bright blue, melon orange, bright red, electric yellow, and chartreuse, among many others. They might form tall showy wands or sit nestled within the center of the vase. Though tropical, there are desert bromeliads as well, and many options for using bromeliads in the mildest, dry-climate gardens, from San Diego to San Francisco. All are excellent container plants and many grow very well in the ground.

Aechmea blanchetiana

COLD TOLERANCE 20–25°F
INTEREST year-round
SIZE 2–4 ft. tall, 1–2 ft. wide
COLOR orange, gold, red
BROMELIAD

This vase-shaped bromeliad, with its brilliant orange or golden yellow blades and pink, red, and salmon flowers, comes to the United States from Brazil. These plants prefer part sun to part shade, and tolerate a wide variety of soils. They grow well in a container, especially one glazed green or oxblood red. While flowers are attractive, grow this bromeliad for its striking leaf colors.

Dyckia

Dyckia is a group of terrestrial bromeliads with rosettes of very spiny-edged, stiff leaves, typically in shades of deep maroon to silver to green to ice green. Vivid orange, red, or gold flowers form on narrow, tall stems. The rosettes do not die after flowering, unlike other bromeliads. Most dyckias are from arid climates in higher-altitude regions of Brazil and Argentina. While small, they usually form pups, eventually growing into broad colonies. They are highly decorative plants that add color and drama to the garden, whether in the ground or in a container. Well-draining soil is important. Keep away from walkways, paths, and other places where people might brush up against the spine edge.

Aechmea blanchetiana

Dyckia 'Black Gold'

COLD TOLERANCE 20–25°F
INTEREST year-round
SIZE 18 in. tall and wide
COLOR purple-black, gold, silver
Bromeliad

Black-purple leaves with an unusual golden yellow flower earn this *Dyckia* its cultivar name. Leaves are frosted looking above and silvery below, with fierce teeth. Grow in full sun.

Dyckia 'Jim's Red'

COLD TOLERANCE 15–20°F
INTEREST summer, fall, year-round
SIZE 1½–2 ft. tall, 3–4 ft. wide
COLOR cabernet, silver, orange
Bromeliad

'Jim's Red' forms a rosette made up of narrow leaves that are bright cabernet-colored above, silvery below. It is very spiny, with a tall flower stalk and orange flowers in summer and fall. Color is best in full sun in coastal and inland gardens.

Dyckia 'Black Gold'

Dyckia 'Jim's Red'

Dyckia 'Precious Metal'

COLD TOLERANCE 20–25°F
INTEREST spring and year-round
SIZE 1½–2 ft. tall, 3–4 ft. wide
COLOR silver, burgundy, red, orange
Bromeliad

This rosette-forming *Dyckia* is made up of narrow burgundy leaves covered in silvery scales that give the leaves a frosted effect. It is *very* spiny, with a tall, narrow, reddish flower stalk with orange flowers in spring.

Dyckia 'Precious Metal'

DRY GARDENING HOW-TO

Here's a secret: low-water gardens are the easiest and most affordable gardens to plant and care for. In this chapter I'll get you started and give you the tools to keep your dry garden thriving. We'll first go over methods for removing your lawn, which is the most difficult part of converting to a low-water landscape. You will also learn how to work with your soil (and not against it); make mounds, swales, and landforms; the best irrigation practices; ways to reorganize waterwise plants; hydrozones; the best planting techniques; establish new plants; manage water over the long term; gray water; the value of mulch; pruning; grooming; fertilization, and more. Once you go low water, you'll never look back!

Removing Your Lawn

Lawn removal is often the first step to creating a colorful, waterwise garden. It might seem a daunting task but there are several methods you can use and some can be used in combination.

First, determine what kind of roots your grass has. The name of the grass is not important; what matters is its roots. Grasses with fine, threadlike roots are easy to kill. Once the blades are dead, the roots die. All of the methods described below work great for these grasses.

Then, there are grasses with thick, fleshy roots that look like the roots that grow out of an old potato. Technically, those are horizontal stems called stolons, but practically they are the gardener's bane. Even after killing the green blades, those roots store energy so new leaves pop up almost overnight. And *do not* rototill these grasses; that breaks the stems up into millions of pieces, each of which can root and produce a new plant. Rototilling multiplies the problem a million-fold. Bermuda grass is the prototypical fat-root grass. All are tough to eradicate. It takes time, it takes patience, and—if you don't want to use chemicals—it takes persistence.

Here are several of the most common methods to try. If one doesn't do the trick, try several in sequence.

Turn the water off If your lawn is entirely fine-root grasses, they will die off without water. Simply turn off the water so the grass dries out and wait.

Dig it out If you have strong arms and a solid back, you can dig your lawn out using a flat-edge shovel. Dig as deep as there are roots. Try not to take out too much dirt, though. The garden needs the dirt and there's no reason to remove it unless it is filled with roots.

If your lawn is very large, rent a sod cutter. Mark spray heads and gas lines with flags so you won't run them over with the sod cutter. Set the blade for as

The Death of Instant Landscapes

Low-water landscapes are the opposite of so-called instant landscapes. Instant landscapes became popular in the late 1970s or early 1980s, primarily for high-end and commercial landscapes, then trickled down to residential landscapes. An instant landscape is planted to look completely grown in the moment the landscape contractor finishes installing the garden. The only way to get that look is to start with large plants, and plant them too close together.

Though an instant landscape looks great when completed, just two or three years later, it becomes an overgrown mess as plants grow into each other and out of scale. The only way to keep them in check is with constant pruning or by "editing"—selectively removing plants there is no longer room for. Both solutions are, frankly, ridiculous. Editing out a half to the third of the plants installed is a waste of money, a waste of labor, and a waste of plants.

I once visited a Southern California garden, for example, where the owner wanted a quick screen for a quarter-mile-long driveway. At her request, her landscape contractor installed two rows of trees planted along the exposed edge of the driveway. One row was California pepper trees (*Schinus molle*) spaced about 6 feet apart. Behind those, 4 feet away, was a row of coast live oaks (*Quercus agrifolia*), also planted about 6 feet apart. In that climate, the natural canopy of each pepper tree would grow to 30–40 feet across at maturity. The oak canopies would eventually reach 40–50 feet across. At that density, each row of trees had eight trees where there should have been just one. And since there were two rows, there were *sixteen* trees in the space for one. Such a waste!

Constant pruning is a maintenance nightmare. It takes enormous numbers of person-hours and generates mounds of greenwaste to be processed into mulch. Mulching is better than sending it to the landfill, but it still requires trucking and processing, both of which consume fossil fuel and generate greenhouse gasses that contribute to global warming.

In addition, plants that are constantly pruned never look natural and never have the chance to show off their true beauty. Most often, these plants get pruned regardless of where they are in their bloom cycle. So often I see a hedge that's been pruned just as the flower buds were about to open—it's such a disappointment.

Instant landscape has run its course. It's time to adopt a new paradigm of waterwise gardens, designed to be beautiful, sustainable, low maintenance, and colorful.

deep as the majority of roots are, then start the engine. It helps to have a strong upper body to control the sod cutter. After you cut out the lawn, irrigate and wait. You may need to repeat the process several times.

Sheet mulch Garden designer Michelle Bellefeuille finds great success sheet mulching over lawn grasses, even Bermuda grass. Sheet mulching is best done in fall so winter rains can keep the area moist. Once the process is completed, the remaining soil is enriched from the organic matter and it will hold water better as well.

1. Mow the grass low. Trench along the edges of the driveway and sidewalks. This is important for keeping the sheet mulch from spilling over onto those surfaces.
2. Irrigate to saturate the soil.
3. Cover the entire surface with thick paper. Newspaper works well, as does brown craft paper that you can typically purchase in rolls. Overlap seams by at least 6 inches.
4. Cover the paper with a 3-inch layer of compost plus another 2–3 inches of bark mulch. Sprinkle in handfuls of worm castings to jumpstart the microbes that will break all the layers down into compost.
5. Make sure to turn off any overhead irrigation systems. Since you covered the pop-up heads, you don't want them to come on and dislodge the sheet mulch.
6. If winter rains are sparse, irrigate with a hose-end sprinkler every so often to keep everything damp. When you water, saturate the layers as deep as possible.
7. Within a few months, the organic matter (including the paper) will have composted in place, killing the grass in the process. Keep an eye out, however, for errant sprouts. When you see them, pull them immediately so they don't take over.
8. Plant by digging holes directly into the sheet mulch, poking holes in any remaining paper. Some people pull the mulch away, plant, and then replace the mulch.

Solarize This is a process that superheats the soil and kills the grass, along with any weed seeds. Unfortunately this also kills some beneficial insects and microbes, but these tend to rebound quickly. Solarization takes six to eight weeks, and is done in the heat of summer. If your thick-root grass has all of its roots near the surface, solarization can do a pretty good job of killing those grasses, too.

1. Mow the grass as short as possible. Remove all rocks, pinecones, and so on. The goal is to have a very smooth surface.
2. Water deeply to saturate the soil; 1–2 feet deep is ideal.
3. Spread 2- to 4-mil *clear* poly sheeting (UV stabilized if possible). *Do not* use black plastic, as it will shade the soil rather than heat

it up. Cover the surface and the grass, plus a foot past the edges, even if that means the plastic covers part of a walkway or sidewalk. The edges will be the coolest area so going beyond the edge ensures even killing. Overlap sheets if you need to.

4. Anchor the edges and seams with bricks, rocks, or two by fours to seal the area under the plastic.
5. Make sure to turn off your irrigation system.
6. Wait. You'll see the grass turn brown within a week or two. After 6–8 weeks, it should all be dead. Recycle the poly sheeting. Irrigate and watch for resprouted grass. Dig out any sprouts you find.

Soil

You don't have to be a soil scientist to create a thriving garden, but knowing a bit about your soil will help you choose the best-suited plants and understand how best to care for them. There are many different soil types across the western United States, from lose sand to dense clay and everything in between. The scale of variation is such that your garden soil may be totally different from your neighbor's garden just a block away.

Soil plays many roles in the garden. It is the medium through which plant roots take up water and nutrients. Soil is where roots and beneficial microbes like fungi and bacteria set up a very complex, interconnected web for exchanging water, carbon, and nutrients. Soil is also where plants are anchored.

Drainage

Soil drainage is a major issue for western gardeners. You often see plants described as preferring "well-drained soil" or being "tolerant of heavy soil." "Well drained" and "heavy" refer to the rate at which water percolates into and flows through soil. Sandy soils are comprised of large particles whose irregular surfaces pack loosely, leaving large (for soil), air-filled spaces between grains. Water filters into and through these pores quickly, so sandy soils are often described as "well draining." Clay soils, on the other hand, are made of much, much tinier particles that pack tightly, leaving far smaller spaces for air and water. Water percolates into and through clay soils very slowly, so they stay wet for much longer. In garden lingo, this slow-draining soil is described as "heavy" soil. Despite being poor draining, clay soils are rich in minerals that plants need. Silt particles are smaller than sand but larger than clay. Drainage rates fall somewhere between the two as well.

Regardless of your soil type, plant roots need oxygen from air as much as they need water. If the soil stays saturated for too long, plant roots drown from lack of oxygen; that is why it is important for soils to dry out in between waterings. That said, every experienced gardener has killed plants by overwatering them, so don't be too hard on yourself if this happens to you.

The bottom line is this: it is more important to know how your soil drains than it is to be able to identify the type of soil in your garden. Run this test in your garden to determine soil drainage:

1. Dig a hole 1–2 feet deep and wide. The deeper and wider, the better.
2. Fill the hole with water and allow it to drain.
3. Fill the hole again and use a ruler to see how long it takes for two inches of water to drain out. Ideally, the soil should drain by at least an inch each hour. If the water drains down two inches in two hours, it is considered to be "well draining." If water is still standing after 24 hours, there might be hardpan or caliche (a layer of soil that has been cemented together) beneath the surface. If you can't break through that layer, plant elsewhere.

My garden is in an old riverbed, so the soil is very sandy. When I do a drainage test, by time I walk to the hose bib, turn the water off, and walk back, the planting hole is completely drained.

Organic Matter

Arid soils are typically low in organic matter. Organic matter serves important functions in soil: it helps fast-draining soils hold water and it helps create pore spaces so heavy soils drain better. As it decomposes and breaks down, organic matter releases nutrients in forms easily taken up by plant roots. In higher-rainfall areas, plants grow big, broad leaves that, when they drop, accumulate in thick layers that compost in place to create deep, rich, organic soils. Arid gardens are totally different. In sunny, arid climates, plants don't generate much biomass, and in the heat, what little biomass there is, breaks down very rapidly. As a result, plants from arid climates have evolved without the need for lots of organic matter. Select the right plants, then, and your garden will thrive, despite the lean soils.

There are other important soil characteristics to be aware of, such as the pH (alkaline soils are typical in the West), salt content (arid climate soils tend to be high in Calcium), and so on. Understanding these characteristics helps you determine how best to proceed in your garden and aids in plant selection.

The best way to understand your soil is to send a sample to a soil lab for analysis. It isn't expensive and it doesn't take long. Be sure to tell the lab what kind of plants you intend to grow (desert natives, Mediterranean natives, and so on). That information helps the soil scientists tailor their analysis and recommendations. A good lab will send you a report that is fairly straightforward to interpret. If you find it confusing, call the lab and ask them to walk you through it.

Mounds, Swales, and Landforms

Soil has yet another, altogether different role. When it comes to garden design, mounded soils and depressions become architectural elements that add dimension to the garden. Think of it this way: a flat garden bed is a flat garden bed, no matter what you plant into it. But a bed with elevation, height, and contour is beautiful even before you add plants.

Bed-sized mounds are sometimes referred to as "Mediterranean mounds." These cover garden beds edge-to-edge or start a few inches in from the edge, rising up to 18 inches or more. Don't be put off by the height. An 18-inch mound can settle to 15 inches or shorter during the first year. It's best to curve mounds through the bed, with undulations in and out, and niches up and down. Feather the edges back to ground level. A bed constructed this way offers many opportunities to show off your beautiful, sculptural, low-water plants.

As an added bonus, if your native soil is slow-draining clay or hard-to-penetrate caliche, a mound or berm of well-draining soil atop the denser soil is a welcome medium for plant roots as they become established. When deeper-rooted plants reach the denser soil, they seem to adapt.

Designer Gabriel Frank puts his driest-growing plants toward the top of the mounds he adds to his designs. Frank sites plants that need a bit more water toward the bottom of the mounds. Even slight differences in available moisture can make all the difference in terms of whether a plant—waterwise or not—struggles or thrives.

Mounds can also form de facto bioswales. Build berms to flank low points or permeable walkways of mulch, decomposed granite, pavers set in sand, and so on. Water from rain or irrigation accumulates at the low points, then penetrates slowly into the soil below.

Irrigation

In a waterwise garden, your goal is to make the most efficient use of water. Irrigation is a major part of that equation. Overhead sprayheads that were long the standard are being replaced by far more efficient irrigation technologies. Rather than a constant shower of water, half of which never makes it to plants, these new technologies release water slowly, are more targeted, and extremely efficient.

In-line drip irrigation is the *most* water-efficient system. While there are many different types of drip irrigation, all release water directly onto the soil at the plant roots, drop by drop, slowly enough to absorb into the soil around plant roots.

In-line drip irrigation is fast becoming the standard for waterwise gardens. This low-pressure technology comes from Israel, where water is extremely precious. At its most basic, in-line drip is flexible poly tubing, a half- or quarter-inch in diameter, with emitters embedded *inside* the tubing. The systems differ a bit from one manufacturer to the next, but they all make several "speeds" of drip. The slowest drippers emit roughly a quarter-gallon per emitter per hour and are designed for clay soils that absorb water very slowly. The fastest emitters release roughly one gallon per emitter per hour and are designed to use with sandy soils that absorb water quickly. The other drip rates are for soils in between those extremes.

Emitters are spaced every 12 inches along the poly tubing. In densely planted garden beds, irrigation lines are laid out on a grid spaced a foot apart. That kind of even spacing ensures that every plant in the bed gets the same amount of water. The flexibility of in-line drip makes it easy to use with odd and irregularly shaped beds, or long and narrow beds like parking strip, as well as in vegetable gardens. For spare plantings, the drip tube is laid out in concentric circles around the base of each plant. Solid poly tubing (without emitters) connects one set of circles to the next, so no water is wasted on bare dirt. All of this tubing sits on top of the soil and gets covered in mulch so it isn't visible but is still easy to find.

There are other approaches to drip irrigation as well. Tucson-area contractor Marcus White of Casa Blanca Horticultural Services establishes plants with drip irrigation. He also contours soils so rainwater collects in areas where plants are located. For his drip systems, White buries half-inch flexible poly tubing 6–8 inches belowground. He pokes individual drip emitters into the line and runs a quarter-inch "spaghetti" tube to each

Mounded soil here directs water toward pathways of decomposed granite, an excellent pathway material that is permeable, so water can percolate down into the soil and stay onsite where it is available to roots of the surrounding trees, shrubs, and other plants.

plant just below the surface. None of the irrigation is visible above ground; all you see is a wet spot next to each plant.

Ollas are an ancient approach to irrigating plants that work well in spare plantings and in vegetable gardens. These porous terra cotta containers are mostly buried, leaving only their necks at the soil surface. Water poured into the olla slowly seeps through the terra cotta and into the surrounding soil to wet plant roots.

Many communities have watering restrictions that typically limit watering days to two per week or fewer and limit run times for the old-fashioned spray systems. They *do not* limit run times for drip, hand watering, or higher technology matched precipitation rate, multi-trajectory spray irrigation. This is a very, very important to understand. High efficiency irrigation systems release water so slowly that it takes a long time to get water deep into the soil. Ten minutes of drip irrigation from a 1-gallon-per-hour emitter releases only a little more than 2½ cups of water per emitter. That is not enough to keep plants healthy.

Hydrozones

Gardeners know to put sun-loving plants in sunny areas of the garden, and shade-loving plants in shady areas of the garden. With waterwise plants, however, we also need to group plants according to their need for water. These kinds of groupings are known as "hydrozones." Hydrozones are irrigated separately, on different irrigation valves, each on its own schedule and running for a different

Drip-irrigation systems conserve water by releasing it directly onto the soil drop by drop.

duration to meet the needs of plants in that zone. Keeping thirsty plants separate from unthirsty plants ensures that every plant gets the amount of water it needs, no more and no less.

The goal with waterwise gardening, then, is to have just a few high-water areas of the garden, located where they'll have the greatest impact. In my garden, the thirsty ornamental plants are in the courtyard by my front door. I see those plants every time I walk into and out of the house. (My vegetable garden, of course, is also filled with thirsty plants, but those plants feed my family.) The rest of the garden requires very little, if any, water.

Planting

When it comes to planting, desert- and Mediterranean-climate plants need tough love. Forget the amendments and fertilizers, just plant directly into native soil for the strongest, healthiest plants. Here's why: well-established plants have wide and deep root systems (this is a generalization, of course, as each plant has a unique root structure, but it basically holds true). When you enrich a planting hole with planting mix or other soil amendments, the roots tend to stay inside that hole where the soil is more hospitable, rather than extending out into the surrounding soils. The result is stunted growth and weak plants.

You have probably seen this phenomenon yourself. Have you ever seen a plant that has been in the ground for a few years, yet hasn't grown as much as expected? If you dig it up to find a surprisingly small root ball, with few roots growing beyond the planting hole, it very likely was planted into a hole of amended soil.

For these types of plants, success is as dependent on the planting technique as on what you mix into the soil. This is my favorite planting method:

1. Well before you start planting, irrigate the entire planting bed so the soil is damp when you plant.
2. Dig a hole as deep as the plant's rootball and about twice as wide. Keep the native soil nearby for refilling the hole later.
3. Rough up the edges of the hole rather than making them smooth.
4. (Optional) Toss in a few handfuls of worm castings. Worm castings contain beneficial microbes that can work symbiotically with plant roots. Use one handful for a 1-gallon pot, four or five for a 15-gallon pot, and so on.
5. Fill the hole with water and let it drain out. If you have clay soil, it might take a while for the hole to drain; sandy soil or decomposed granite drain quickly.
6. Water the plant in its container, until water runs out the holes in the bottom.
7. Once the planting hole drains, remove the plant from its container and rough up the root ball. Some

Contractor Marcus White installs drip irrigation below ground in Tucson-area gardens. When the water runs, all that is visible are these wet spots at the base of each plant.

How to Recognize Waterwise Plants

In the nursery, we are drawn to the plants that look most beautiful at just that moment, but those are not necessarily the best plants for your dry-growing garden. It would be best to shop for plants the way you shop for groceries: with a well-thought-out shopping list. But really, who does that? So when you head to the nursery, at least know how to recognize waterwise plants. These plants share a set of basic characteristics that help them gather, hold, and conserve water. Look for:

Gray or silver leaves These colorations protect leaves by reflecting intense sunlight that characterizes hot, arid environments.

Small and/or narrow leaved plants Plants lose water through pores (stomates) in their leaves. The fewer stomates, the less water loss, and the smaller or narrower the leaf, the fewer stomates.

Fuzzy plants What we perceive as "fuzz" is usually tiny, short, dense hairs that cover leaf and sometimes stem surfaces. The fuzz serves two practical functions. First, like shadecloth, it protects leaves from the sun. Second, it serves as a vapor barrier, slowing the rate of water loss from the leaves.

Plants that hold their leaves upright and close to the stem Horizontal leaves endure the full brunt of bright sunlight. Those held at an angle and close to the stem get as much sun as they need but not so much as to cause damage.

Leathery leaves Since tough, leathery leaves don't hold much water, they don't have much to lose either.

Succulents The roots, stems, and leaves of succulent plants have evolved to take up and hold water when it is available, and store it to use when conditions are dry.

Spine-covered plants Spines do more than protect plants from critters, they also offer shade. Water that condenses onto spine surfaces trickles down to the soil for plant roots to use, too.

Waxy leaves All leaves have a fine layer of wax covering, but some plants in hot, arid conditions produce much more wax that helps to seal water in. Touch a leaf and the white, powdery wax will come off on your fingers.

Be sure to do your homework and confirm that a plant is waterwise before you purchase it.

people even remove all the dirt from around the roots before they plant so the plant goes into entirely native soil in the hole. Note: *Bougainvillea* is an exception. Try to avoid disturbing its roots.
8. Place the plant into the hole, taking care not to bend or break any of the roots.
9. Backfill the hole with native soil. Gently tamp the soil and firm it around the root ball as you go along.
10. As you finish with each plant, trickle water onto the surface to settle the soil around the roots. Allow the water to saturate the entire rootball and soil in the hole. This process eliminates air pockets where roots might dry out and cause the plant to fail.

Finally, be sure to keep plants well irrigated from the time they are planted until they are established—damp, but not wet. This is critical to their success.

When to Plant

Deciding when to plant is as much an art as it is a science. In Mediterranean regions of California, the best times to plant are fall, winter, and early spring. Once the weather cools, the soil is still warm but the heat eases and rains are (hopefully) soon to arrive. Warm soil helps encourage roots to grow, cool air is less stressful on plants at transplant time, and once it starts to rain everything stays damp (not wet) until spring. In valleys and inland areas, finish planting at the end of April or early May so plants start to establish before the weather gets hot. Right along the coast, and in northern regions, planting ends a bit later in the year. We tend to avoid planting in the hottest months; it can be done, but it takes more attention and more irrigation to get plants established. And, in the heat, there is a higher rate of mortality.

Desert planting also revolves around rainfall, with deference to heat and cold. In Phoenix and Tucson, for example, best planting times are from the end of the summer monsoon rains in late September through November when the weather grows cold, then from February through early April, after which the weather is too hot.

In New Mexico, the best planting times depend on location, elevation, and on the particular plants. Some plants are best installed in fall so they can harden off before winter cold. Other plants are best planted in spring to establish before it gets too hot. Some can handle summer planting, others winter. Such variation can complicate a garden's installation, so the pros plant in phases. A garden's preparation and infrastructure—such as grading, hardscape, and irrigation—are done at one time. Plants are installed based on how well they are suited to the season. Mark the locations of those to be planted later, then return to plant them in their best seasons.

Establishing New Plants

No plant is waterwise the moment it goes into the ground, even if it will eventually survive solely on rainfall or just the occasional irrigation. All new plants need to be irrigated regularly while they are becoming established. Establishment takes one to three years, during which time each plant develops the extensive root system it needs to take up water and nutrients from the soil.

In that establishment period, irrigate often enough to keep the rootball damp—not wet. That's an important distinction since roots can drown in soil that is too wet. Stick your finger down into the soil to see if it is dry enough to warrant watering. If you prefer a more technical approach, use a soil probe to pull up a core of soil. If the core shows dry soil several inches below the surface, it's time to irrigate.

Each plant becomes established on its own timetable, but as a general rule, the smaller the plant, the shorter the establishment period. Where perennials can become established in a season, a tree could take up to three years.

There's a lot to learn from simply observing your plants, too. Drooping leaves mean that leaves are losing water faster than roots can pull it out of the soil. If leaves are droopy at the end of the day, the plant will likely recover overnight. But if the leaves are droopy in the morning, it's time to water. A plant that stays too dry for too long will die.

Don't skimp on the water during establishment. Well-established plants make for strong, drought-resilient gardens. The ultimate goal is a garden that lives on rainfall or just infrequent, deep irrigation.

Managing Water

Once plants are established, it's time to cut back on irrigation. "Cutting back" in this case means watering less often but without changing the time each irrigation zone runs. So if a garden bed was watered once a week for 45 minutes during establishment, continue to water it for 45 minutes, whether once a month or once a quarter.

There are two ways to manage irrigation in an established garden. You can turn each valve on and off manually, or you can install an irrigation controller. The old-style irrigation controllers that are set to run for a set number of minutes on particular days have been replaced by "smart" irrigation controllers. Smart controllers monitor the weather while also using information about your garden to "decide" how often and when to run each irrigation zone. You program in your zip code and then enter information about the type of soil, type of plants, type of irrigation, and so on for each irrigation zone.

Smart irrigation controllers can be set to water on specific days—or to *not* water on specific days. If you live in a community where irrigation is limited to Mondays and Fridays, or only on even-numbered or odd-numbered days, for example, these controllers accommodate those schedules.

OPPOSITE: Here, a dry streambed acts as a bioswale to capture water as it runs off the slope and across the property. The water then slowly infiltrates the soil. The goal is to keep water onsite and available to plants.

Gray Water

Gray water is water left over from sinks, showers, tubs, and washing machines that would otherwise go into the sewer or septic system. Repurposing water from your household is popular for good reason, but in the garden I recommend it only for thirsty fruit trees—not for waterwise plants. I am concerned about managing the amount of gray water diverted into my waterwise garden. My family of three adults does five or six loads of laundry each week (we did many more when there were four of us and the children were small). We have a laundry-to-landscape system, so every time the washing machine runs, the garden gets irrigated. Five or six loads of laundry generate too much water, and too often for waterwise plants. My fruit trees, however, are thriving. If my system were more flexible and we could move hoses from plant to plant between loads, I might feel differently. If you do use a similar system in your garden, be careful to use the right laundry soaps and add a diversion valve in case you are washing something whose effluent you don't want to go into the garden, or are using bleach or disinfectant.

As much as we like the idea of technology that takes care of itself, however, there really is no such thing. Check your irrigation controller regularly to make sure all the zones are running as you expect. While you are checking, look at the controller's data log. You'll find a running record of your garden's high and low temperatures, as well as the precipitation. It's like having a simple weather station in your own backyard.

In the rainy season, the garden's irrigation can be turned off completely—after establishment, of course. In normal rainfall years, I turn the irrigation off as the rains start in November in my San Diego–area garden. It stays off, save for periodic watering, until March or April.

Another aspect of water management is managing the water that hits the ground. The goal here is to keep as much water on-site as possible. Because the best place to store water is in the soil, bioswales (intentional depressions where water can sit as it percolates down into the soil) are the first choice for water collection.

You can also capture rainwater in cisterns or rain barrels and use it over time. Rain barrels, however, are more challenging in Mediterranean climates where all the rain falls at once, so you have to store the water from fall until spring. Still, there is a case to be made for capturing and using as much rainfall as you can.

Mulch

Mulch is a critical element in waterwise gardens. Whether organic (plant based) or inorganic (rock, decomposed granite, or other nonliving material), a 3- to 4-inch layer of mulch on top of the soil insulates the soil to slow evaporation and keep soil moist. By shading the soil, mulch deprives weed seeds of the sunlight they need to develop and grow. Weed seeds that land in mulch and germinate tend to root loosely so they are extremely easy to pull.

Organic mulches are made from leaves, plant prunings, wood chips, pine needles, kitchen scraps, and other kinds of plant materials. These kinds of mulches absorb and hold water, much like a sponge. As the organic matter breaks down, it slowly releases nutrients into the soil and helps improve soil texture. Tiny critters and microbes play a role in the decomposition process while also helping plant roots obtain water and nutrients. Organic mulches also help moderate soil temperature, keeping it cooler in summer and warmer in winter.

Broad-leaved, nonsucculent plants do very well with organic mulch, but succulents are better mulched with inorganic materials. Inorganic mulch—most typically rock or gravel—has some additional advantages. Rock keeps soil temperature moderated in times of extreme heat and extreme cold. When water vapor in the air condenses onto rock, droplets trickle down to the soil, adding to available moisture for plant roots.

The value of rocks as mulch was well known to ancient civilizations, including the Southwest's ancient Anasazi peoples, who used rock and cobble to mulch corn, cotton, agave, and other crops.

Today's dry gardens benefit from cobble mulch too. Not long ago, I designed a garden for a home whose native soil was large cobbles suspended in clay. The workers cursed as every shovel into the earth hit a rock. They gathered the rocks and piled them on the driveway to discard, but to me those rocks were gold. I had them use the cobbles to mulch as much of the garden as they would cover, and used organic mulch for the rest. Six months later, the garden was thriving, but the plants mulched with cobble were twice the size as the same plants mulched with organic matter.

Whatever kind of mulch you chose, always keep it several inches away from the crown of the plant, its trunks, and main stems. It isn't a problem for plants to grow over and onto mulch, but mulching over a crown, or against a stem or trunk, can cause rot. Your goal is to have a bare area right around the base of plant.

Should mulch get trapped between the leaves of agaves, aloes, bromeliads, phormiums, kangaroo paws, or other vase-shaped plants, brush or vacuum it out, both to prevent rot and for the aesthetics.

New gardeners are often under the mistaken impression that they need weed cloth beneath mulch. Weed cloth is *totally* unnecessary—in fact it prevents

Organic mulch improves soil texture and gradually releases nutrients as the organic matter breaks down.

organic mulch from doing the important job of improving soil and supporting soil microbes as it breaks down. At the same time, weed cloth does nothing to stop weeds from landing in mulch and germinating. Under rock mulch, weed cloth is simply unnecessary.

Pruning

Put an 8-foot-wide plant in a 5-foot-wide space and you'll constantly be pruning. Put a 4- or 5-foot-wide plant in that same spot and never prune again. Which makes the most sense? I encourage gardeners to create gardens that don't require pruning. You can always *choose* to prune of course, to create a particular effect like espalier or an arch, or if you find pruning therapeutic (I admit to that), but putting a plant into a space that is too small means it will require maintenance forever—that's simply silly.

Instead, go through your garden plants once every three to six months to cut out dead wood, remove crossing or rubbing branches, and redirect any branches that happen to be poking into a walkway or blocking a view.

When it comes to flowering plants, wait until *after* flowering finishes, then prune. If you prune too early, you'll cut off all the flower buds, then wonder why the plant isn't blooming or fruiting.

If you enjoy cut flowers, grow plants in the family Proteaceae, such as proteas, grevilleas, pincushions, and conebushes. Wherever you cut a branch, four or five or more new branches soon sprout at the cut point. Since flowers form at the tips of the branches, the more you prune, the more branches, and the more flowers the plants make.

Grooming

Bulbs like *Watsonia*, grasses like *Muhlenbergia*, and other plants like *Kniphofia* and *Carex* grow long, narrow blades that benefit from an occasional grooming to remove dead leaves.

Deciduous grasses go dormant, mostly in winter, then sprout fresh leaves in spring. Leave the brown blades through winter to enjoy their strong structures and sculpture. Or, cut dormant grasses back to 4- to 8-inch-tall mounds (taller for taller grasses) and wait for fresh foliage to resprout when the weather warms.

Cutting a large clump of tall grass is more challenging than it looks. Here's a trick to make it much easier: wrap a long piece of twine around the outside of the clump to encircle it, about a foot above the ground. Pull the twine so the blades are gathered in a tight bundle. Once the twine is holding the blades securely, simply cut across the blades with pruning shears *below* the level of the twine.

Evergreen grasses are better groomed than cut back. Comb out dead blades and leave green ones in place.

Fallen leaves and flower petals are mulch. They are grooming issues only when they fall onto pathways, pebbles, dry streambeds, or fill the crevices in upright, vase-shaped plants. Rake them up as best you can. For debris too small for the rake or a hand broom, try a shop vac. That's not a joke. The large California pepper tree in my front garden drops a huge volume of tiny leaves and minute yellow flowers (it's a male tree). As much as I appreciate the tree's beauty and its ability to self-mulch, that debris will smother the agaves, aloes, bromeliads, and other plants under its canopy. It is nearly impossible to clean out by hand. So, three or four times a year, we use our shop vac to suck it all up. It takes very little time and what gets sucked out goes into the compost pile. The plants look clean and beautiful once again.

Fertilizing

Generally speaking, forget about fertilizer; low-water plants grow perfectly well on a low-nutrient diet. Organic mulches (plant-based mulches) provide whatever nutrition the plants need. In fact, that's how these plants evolved. Save time, save money, and don't fertilize.

That said, if you feel you absolutely *must* fertilize, do so with caution. There is some evidence that high concentrations of fertilizers suppress the activity of beneficial soil microbes that help keep plant roots healthy. Also, fertilizers tend to push fast growth, which tends to be vulnerable to pests. And fast growth is the opposite of low maintenance.

For plants in the family Proteaceae—*Protea, Grevillea, Leucospermum, Leucadendron,* and so on—use *only* low-phosphorus fertilizers if you use any fertilizer at all. Phosphorus is the "P" in NPK, the three macronutrients that are the major components of general fertilizers. Every fertilizer label has an "NPK" rating that describes the ratio of nitrogen (N) to phosphorus (P) to potassium (K). A 10:10:10 fertilizer, for example, has equal parts nitrogen, phosphorus, and potassium and is known as a "balanced" fertilizer. In other words, that is *not* the fertilizer to use for these plants. Instead, look for a formulation where the middle number is the lowest of the three, then follow label directions.

And if you must fertilize, choose organic fertilizers over conventional ones. Organic fertilizers are derived from minerals, plants, and animals. They are in various states of decomposition and will continue to break down slowly by soil microorganisms and weathering. That makes them naturally "slow-release" fertilizers and reduces the potential for burning plants or damaging beneficial soil microbes.

In almost all cases, however, patience is the best fertilizer.

RESOURCES

Suggested Reading

Calhoun, Scott. 2012. *The Gardener's Guide to Cactus: The 100 Best Paddles, Barrels, Columns, and Globes*. Portland, OR: Timber Press.

Irish, Mary. 2002. *Arizona Gardeners Guide*. Minneapolis, MN: Cool Springs Press.

———. 2008. *Trees and Shrubs for the Southwest: Woody Plants for Arid Gardens*. Portland, OR: Timber Press.

Irish, Mary F., and Gary Irish. 2000. *Agave, Yuccas, and Related Plants: A Gardeners Guide*. Portland, OR: Timber Press.

Perry, Robert C. 2010. *Landscape Plants for California Gardens*. Los Angeles, CA: Land Design Publishing.

Phillips, Judith. 2005. *New Mexico Gardeners Guide*. Minneapolis, MN: Cool Springs Press.

———. 2015. *Growing the Southwest Garden*. Portland, OR: Timber Press.

Sterman, Nan. 2007. *California Gardeners Guide*, vol. II. Minneapolis, MN: Cool Springs Press.

Online Resources

A Growing Passion (TV show hosted by Nan Sterman)
AGrowingPassion.com
Watch, in particular, episodes 102 ("California Native Grown"), 104 ("Waterwise and Wonderful"), 105 ("Cycle and Recycle"), 202 ("Chaparral, the Elfin Forest"), 306 ("How Water Flows"), 404 ("Bye Bye Grass"), and 405 ("After the Lawn is Gone")

Alliance for Water Efficiency
allianceforwaterefficiency.org

Artyfactory (on color theory)
artyfactory.com/color_theory/color_theory.htm

California Native Plant Society
calscape.org

Chihuahuan Desert Education Coalition
chihuahuandesert.org

Desert Tropicals
desert-tropicals.com

EPA Watersense
epa.gov/watersense

Gardening in Arizona
GardeningInArizona.com

Metropolitan Water District of Southern California
bewaterwise.com

National Drought Mitigation Center
drought.unl.edu

Native Plant Information Network, Ladybird Johnson Wildflower Center
wildflower.org/plants

New Mexico State College of Agricultural, Consumer, and Environmental Sciences, Low Water-Use Landscape Plants for the Southwest
aces.nmsu.edu/pes/lowwaterplants

SelecTree: A Tree Selection Guide
selectree.calpoly.edu

Resources

Southern Nevada Water Authority
snwa.com/apps/plant/index.cfml

Tucson Cactus and Succulent Society
tucsoncactus.org

University of Arizona Cooperative
    Extension, Pima County
extension.arizona.edu/pima

USDA Plant Guides and Database
plants.usda.gov

US Drought Monitor
droughtmonitor.unl.edu

Water Footprint Network
waterfootprint.org

Retail and Online Nurseries Specializing in Waterwise Plants for the West and Southwest

Arizona

B&B Cactus Farm
Tucson
bandbcactus.com

Civano Nursery
Tucson
civanonursery.com

Desert Survivors
Tucson
desertsurvivors.org/plant-nursery.html

Plants for the Southwest and Living Stones
    Nursery (mail order)
Tucson
plantsforthesouthwest.com and
    lithops.com

Summer Winds Nursery
Phoenix
summerwindsnursery.com/az/home

Treeland Nursery
Mesa
treeland.com

Verde Valley Nursery
Fountain Hills
verdevalleynursery.com

Whitfill Nursery
Phoenix
whitfillnursery.com

California

Annie's Annuals & Perennials
Richmond (and online)
anniesannuals.com

Australian Native Plants
Casitas Springs
australianplants.com

California Flora Nursery
Fulton
calfloranursery.com

East Bay Wilds
Oakland
eastbaywilds.com

Flora Grubb Gardens
San Francisco
floragrubb.com

Grow Native Nursery at Rancho
    Santa Ana Botanic Garden
Claremont
rsabg.org/grow-native-nursery

Las Pilitas Nursery
Santa Margarita
laspilitas.com

Oasis Water Efficient Gardens
Escondido
oasis-plants.com

Serra Gardens
Fallbrook (and online)

shop.cacti.com/serra-gardens

Sierra Azul Nursery & Gardens
Watsonville
sierraazul.com

Succulent Gardens
Castroville (and online)
sgplants.com

Theodore Payne Foundation for Wild
    Flowers and Native Plants
Sun Valley
theodorepayne.org

Tree of Life Nursery
San Juan Capistrano
californianativeplants.com

Waterwise Botanicals
Bonsall
waterwisebotanicals.com

Yerba Buena Nursery
Half Moon Bay
yerbabuenanursery.com

New Mexico
High Country Gardens (online only)
highcountrygardens.com

Plants of the Southwest
Santa Fe and Albuquerque (and online)
plantsofthesouthwest.com

Wholesale Nurseries

Australian Outback Plantation
Tonopah, Arizona
australianoutbackplants.com

C&S Nursury
Los Angeles, California
csnursery.com

East West Trees
Fallbrook, California
eastwesttrees.com

El Nativo Growers
Azusa, California
elnativogrowers.com

Monterey Bay Nursery
Royal Oaks, California
montereybaynursery.com

Mountain States Wholesale Nursery
Glendale, Arizona
mswn.com

Native Sons Nursery
Arroyo Grande, California
nativeson.com

San Marcos Growers
Santa Barbara, California
smgrowers.com

Suncrest Nurseries
Watsonville, California
suncrestnurseries.com

Public Gardens with Exemplary Displays of Waterwise Plants

Arizona
Arizona-Sonora Desert Museum
Tucson
desertmusuem.org

Boyce Thompson Arboretum
Superior
arboretum.ag.arizona.edu

Desert Botanical Garden
Phoenix
dbg.org

Tohono Chul Gardens, Galleries, Bistro
Tucson
tohonochulpark.org

Tucson Botanical Gardens
Tucson
tucsonbotanical.org

Resources

University of Arizona Campus Arboretum
Tucson
arboretum.arizona.edu

California

Ganna Walska Lotusland
Santa Barbara
lotusland.org

Huntington Library, Art Collection, and Gardens
San Marino
huntington.org

John R. Rodman Arboretum, Pitzer College
Claremont
pitweb.pitzer.edu/arboretum

Leaning Pine Arboretum, California Polytechnic State University
San Luis Obispo
leaningpinearboretum.calpoly.edu

Living Desert Zoo and Gardens
Palm Desert
livingdesert.org

Los Angeles County Arboretum and Botanic Garden
Arcadia, California
arboretum.org

Natural History Museum of Los Angeles County Nature Gardens
Los Angeles
nhm.org/nature/visit/nature-gardens

Rancho Santa Ana Botanic Garden
Claremont
rsabg.org

Ruth Bancroft Gardens
Walnut Creek
ruthbancroftgarden.org

San Diego Botanic Garden
Encinitas
sdbgarden.org

Santa Barbara Botanic Garden
Santa Barbara
sbbg.org

University of California Botanical Garden at Berkeley
Berkeley
botanicalgarden.berkeley.edu

University of California Santa Cruz Arboretum
Santa Cruz
arboretum.ucsc.edu

Water Conservation Garden
El Cajon
thegarden.org

New Mexico

Albuquerque BioPark Botanical Garden
Albuquerque
cabq.gov/culturalservices/biopark/garden

Santa Fe Botanical Garden
Sante Fe
santafebotanicalgarden.org

Utah

Red Butte Garden, University of Utah
Salt Lake City
redbuttegarden.org

Water Conservation Garden Park
West Jordan
conservationgardenpark.org

PHOTOGRAPHY CREDITS

All photos are by the author, except for the following:

Anthony Tesselaar USA, page 266

Randy Baldwin, page 192

Sarah Bryant, Tree of Life Nursery, pages 222, 223 (left)

Laura Camp, Tree of Life Nursery, pages 178 (right), 195, 198, 209 (left), 210 (left), 223 (right)

Susan E. Degginger / Alamy Stock Photo, page 226 (top)

Flickr / Andrey Zharkikh, used under a Creative Commons 2.0 generic license, page 274 (right)

Flickr / Jean-Michel Moullec, used under a Creative Commons 2.0 generic license, page 220 (right)

Flickr / Peter G. Veilleux, page 208

Flickr / Rob Klotz, page 197 (right)

GAP Photos / Rob Whitworth, page 176

Garden World Images Ltd. / Alamy Stock Photo, pages 210 (right), 282 (left)

Kelly Griffin, page 183 (right)

HighCountryGardens.com, pages 252, 280

Saxon Holt, page 282 (right)

Horticopia / Dr. Edward F. Gilman, page 196 (left)

Joshua McCullough, page 243

Courtesy of Monrovia, page 221 (left)

Mountain States Wholesale, pages 224 (right), 262, 263 (left), 268

Shutterstock.com / guentermanaus, page 216 (right)

Sunset Western Garden Collection, page 215

Wikimedia / Michael Wolf, used under a Creative Commons Attribution–Share Alike 3.0 Unported license, page 190 (left)

Wikimedia / Rosina Peixoto, used under a Creative Commons Attribution–Share Alike 3.0 Unported license, page 212 (right)

Wikimedia / Stan Shebs, used under a Creative Commons Attribution–Share Alike 3.0 Unported license, page 240

Wikimedia / Stickpen, public domain, page 196 (right)

INDEX

Abutilon palmeri, 57
Acacia, 150
Acacia boormanii, 112
Acacia greggii, 148
Acacia pendula, 73, 150
Acacia schaffneri, 150
Acacia stenophylla, 150
Acca sellowiana, 112
Achillea, 176
    'Hella Glashoff', 176
    'Moonshine', 77, 80, 81, 154, 179
Achillea filipendula, 176
Achillea millefolium, 71, 72
    var. *californica*, 177
    'Paprika', 61, 67, 178
    var. *rosea* 'Island Pink', 178
Adenanthos cuneatus 'Coral Drift', 27
Adenanthos sericea, 77
Aechmea blanchetiana, 280
Aeonium, 29, 103, 173, 179
    'Cyclops', 81, 180
    'Kiwi', 19, 89, 180
    'Sunburst', 53, 57, 122, 140, 146, 181
    'Velour', 32
    'Zwartkop', 33, 53, 57, 74, 77, 81, 114, 119, 122, 140, 146, 182
Aeonium nobile, 142, 181
Aeonium urbicum, 85
Aethionema schistosum, 132, 135
African daisy, 22, 50, 168, 198
African starfish flower, 161–162
agave, 105, 119, 154, 299
Agave, 182, 253
    'Blue Elf', 105
    'Blue Flame', 87, 100, 103, 184
    'Blue Glow', 127, 128
    'Joe Hoak', 57, 121, 122, 185
    'Kissho Kan', 185
Agave americana var. *mediopicta* 'Alba', 85, 183
Agave applanata 'Cream Spike', 183
Agave attenuata, 37, 77, 81, 85, 173
    'Kara's Stripes', 85, 86, 144
    'Nova', 85, 184
Agave guiengola, 50, 51
Agave lophantha 'Quadricolor', 100
Agave parryi, 37
    var. *truncata*, 86
Agave potatorum 'Variegata', 185
Agave vilmoriniana 'Stained Glass', 98, 103, 105, 122, 140, 146, 186
Agonis flexuosa 'After Dark', 100
airplane plant, 216
Alluaudia procera, 121
aloe, 119, 170, 299
Aloe, 186
    'Blue Elf', 119, 122, 187
    'David Verity', 57, 188
    'Jacob's Ladder', 50
Aloe cameronii, 57, 100, 105, 174, 187
Aloe camperi, 169
Aloe ferox, 170, 188
Aloe marlothii, 56
Aloe plicatilis, 42, 127, 128, 142
Aloe rubroviolacea, 57, 167, 169, 174, 189
Aloe saponaria, 85
Aloe speciosa, 102, 105, 189
Aloe spicata, 190
Aloe striata, 23, 85, 87
Aloe vanbalenii, 144

Index

Alstroemeria, 110
    'Third Harmonic', 168, 171, 174, 190–191
Anderson, Patrick, 49–57
Anemanthele lessoniana, 98, 101
Anigozanthos, 91, 95, 191
    'Big Red', 42, 118, 122, 174, 191
    'Bush Dawn', 89, 121, 122, 192
    'Bush Tango', 121
Arabian aloe, 189
Arbutus 'Marina', 114, 192–193
Arctostaphylos, 194, 236
    'John Dourley', 67, 196
    'Lester Rowntree', 65, 66, 67, 197
    'Sunset', 61, 198
Arctostaphylos densiflora 'Sentinel', 67, 194–195
Arctostaphylos edmundsii var. *parvifolia*, 70, 196
Arctostaphylos pungens, 197
Arctotis, 50, 198
    'Big Magenta', 168, 170
    The Ravers 'Pink Sugar', 26, 168, 174, 199
    The Ravers 'Pumpkin Pie', 22, 67, 199
Aristolochia gigantea, 44, 80
Arizona, 176
Arizona queen of the night, 160
Artemisia californica, 71
ascot rainbow spurge, 230
Asparagus densiflorus 'Myers', 103
Asteriscus maritimus, 91
Astrophytum myriostigma, 161
autumn sage, 150, 268

backdrop, 15–17
Baileya multiradiata, 151
Baja fairy duster, 167
barberry, 95
Barragán, Luis, 84
barrel cactus, 232

Bassage, Jay, 115–122
beaked yucca, 102
bearded iris, 238
bear grass, 132
Beaucarnea, 44
Beaucarnea guatemalensis, 119, 121
Beaucarnea pliabilis, 103
bees, 262, 270
bee's bliss sage, 269
Bellefeuille, Michelle, 41, 58–65, 286
Berberis thunbergii var. *atropurpurea* 'Crimson Pygmy', 93–95
Berlandiera lyrata, 132, 135, 136, 138, 150, 153, 200
Bermuda grass, 286
Beschorneria yuccoides, 86, 200
Bignonia capreolata 'Tangerine Beauty', 135, 138, 201
big red kangaroo paw, 191
Billbergia, 126
billy buttons, 80
bioswales, 104, 105, 289
bird of paradise bush, 50, 147
birds, 111, 151
bishop cap, 161
blackbird spurge, 228
black spine prickly pear, 258
blanket flower, 43, 100, 235
blue chalk fingers, 43, 85, 119, 128, 143, 144, 170, 274
blue chalksticks, 53, 87
blue columnar cactus, 49, 264
blue foxtail agave, 184
blue hesper palm, 100, 203
blue palo verde, 147
blue sedge, 105
bottle tree, 119
Bougainvillea, 201
    'Bengal Orange', 26, 168, 171, 174, 202
    'California Gold', 126
    'Tahitian Dawn', 126

Bougainvillea brasiliensis, 102, 103, 105, 114, 202
Brachychiton rupestris, 119
Brahea armata, 100, 203
brittlebush, 24
bromeliads, 126, 280–283, 299
bronze mat manzanita, 70, 196
Bryophyllum fedtschenkoi, 240
Buchloe dactyloides 'UC Verde', 119
buckwheat, 222–224
Buddleia marrubifolia, 154, 203
buffalo grass, 119
bugle lily, 170, 173
burgundy cordyline, 215
bush dawn kangaroo paw, 192
bush germander, 127
bush penstemon, 262
butterflies, 176, 200, 203, 262, 268

cabbage tree, 119
Caesalpinia gilliesii, 147
Caldwell, Chris, 136
Calhoun, Scott, 20, 157–162
Calibanus hookerii, 161
California, 176
California buckwheat, 223
California bush sunflower, 71
California flannel bush, 165
California fuchsia, 63–65, 222
California grape, 20
California lilac, 112, 165, 206, 208
California pepper tree, 50, 285, 302
California poppy, 61, 170, 225
California sagebrush, 71
Calliandra californica, 167
Calliandra 'Sierra Starr', 163, 204
Calliandra surinamensis, 53
Calylophus berlandieri subsp. *berlandieri*, 205

Calylophus drummondianus, 26, 99, 101, 105, 167, 168, 171, 174, 205, 269
Calylophus drummondii, 205
camphor, 98
candellia, 227
cape aloe, 188
Cape honeysuckle, 107
Carex, 46, 301
Carex glauca, 105
Caribbean copper plant, 119, 229
Carl, Debra, 28
Carnegiea gigantea, 160
cassia, 206–207
Cassia, 206
Cassia artemesoides, 206, 207
Cassia nemophila, 207
Cassia phyllodinea, 207
catclaw acacia, 148
catmint, 93
Ceanothus, 165, 206, 208–210, 236
    'Centennial', 67, 208
    'Concha', 130, 209
    'Dark Star', 70, 72, 209
    'Frosty Blue', 70, 72, 210
Ceanothus gloriosus var. *exaltatus* 'Emily Brown', 70, 72, 210
Ceanothus griseus 'Diamond Heights', 112
centennial California lilac, 208
Cercis canadensis 'Forest Pansy', 20
Cestrum 'Newellii', 63, 66
chalk liveforever, 71, 144
changeling monkey flower, 256
chaparral sage, 270
Chilopsis linearis, 20
Chinese privet, 123
chocolate flower, 132, 135, 136, 150–151, 200
Chrysactinia mexicana, 150, 153
Cistus, 135, 138, 211
Cistus ×*purpureus*, 135, 211

Cistus ×*skanbergii*, 67, 212
Clarkia unguiculata, 70
clay soil, 287
Cleveland sage, 70, 150, 270
Clivia, 128, 130, 212
Clivia nobilis, 130
coastal wooly bush, 77
coast live oak, 285
cobble mulch, 299
Cohen, Murray, 68
Coleman, Marcia, 58–63
Coleonema 'Sunset Gold', 44, 77, 80
color
    architecture, 31
    backdrop, 15–17
    designing with, 14
    details, 35
    hot, 14
    pottery, 31–35
    schemes, 17–30
"color intersections", 170
concha California lilac, 209
conebush, 29, 85, 100, 124, 167, 246–248, 301
containers, 86–87, 89, 92, 143–144, 152, 160–161
contrasting color schemes, 24–25
Convolvulus mauritanicus, 44
cool color schemes, 25
coppertone stonecrop, 50, 77, 85, 273
coral aloe, 87
cordyline, 46
Cordyline, 213
    'Design-A-Line Burgundy', 98, 101, 105, 215
    'Festival Grass', 27, 80
    'Red Sensation', 168, 173
Cordyline australis
    'Red Star', 84
    'Torbay Dazzler', 98, 101, 105, 213

Coreopsis auriculata 'Nana', 60, 61
Cotinus coggygria, 110, 114, 167, 168, 170–172, 174, 214
    'Purple Robe', 214
    'Royal Purple', 130, 214
Cotyledon orbiculata var. *oblonga* 'Flavida', 77
crape myrtle, 243
Craspedia globosa, 27, 80
Crassula, 215
Crassula capitella 'Campfire', 89, 143, 146, 216
Crassulacean acid metabolism (CAM), 240
Crassula falcata, 216
Crassula ovata 'Hummel's Sunset', 122, 146, 217
creeping fig, 107
Crocosmia, 170
Crocus vernus 'Remembrance', 61
cross vine, 201
crown of thorns, 74, 119, 230
Cussonia paniculata, 119, 121
Cyperus papyrus, 100

Dalea capitata, 163, 217
Damianita daisy, 150
dark star California lilac, 209
Dasylirion acrotrichum, 147
Dasylirion longissimum, 84
Dawe's aloe, 50
daylily, 91
Dean, Leslie K., 90–95
Dean, Nick, 68–70
Dean Design Landscape Architecture, 90
deciduous grasses, 301
desert bluebell, 160
desert cassia, 207
desert mallow, 275
desert marigold, 151

Index

desert museum palo verde, 166–167, 259
desert ruellia, 268
desert sage, 136, 270
desert spoon, 147
desert willow, 20
desert zinnia, 135, 280
design
    for plant structure and texture, 38–47
    with color, 14–37
details, 35
Dianella caerulea 'Cassa Blue', 27
dittany of Crete, 43, 128
Dodonaea viscosa 'Purpurea', 80
Dracaen, 213
drainage, 287–289
drip irrigation, 60, 87, 94, 137–138, 144, 289–291
Dudleya pulverulenta, 71, 144
dwarf goldenrod, 274
dwarf honeybush, 80
dwarf tickseed, 61
Dyckia, 281–283
    'Black Gold', 74, 75, 81, 282
    'Jim's Red', 100, 102, 105, 282
    'Precious Metal', 100, 105, 283

eastern redbud, 20
ebony conebush, 247
echeveria, 218–221
Echeveria, 43, 218
    'Afterglow', 57, 108, 114, 218
    'Etna', 143, 146, 220
    'Mauna Loa', 142, 143
    'Perle von Nürnberg', 33
Echeveria agavoides 'Lipstick', 142, 146, 219
Echeveria elegans, 57, 146, 219
Echeveria gigantea, 77, 81, 220
Echeveria ×imbricata, 221
Echinocactus grusonii, 50, 57, 170, 221

Echium wildpretii 'Tower of Jewels', 100
Ella Nelson's naked buckwheat, 224
Emily Brown California lilac, 210
Encelia californica, 71
Encelia farinosa, 24
Epilobium californica, 222
Epilobium canum, 221
    'Schieffelin's Choice', 41, 63, 67, 222
Eriogonum, 222
Eriogonum fasciculatum, 223
Eriogonum giganteum, 129
Eriogonum grande var. *rubescens*, 63, 66, 67, 72, 95, 223
Eriogonum nudum 'Ella Nelson's Yellow', 65, 66, 67, 224
Eriogonum wrightii, 224
eryngium, 18
escarpment live oak, 133
Eschscholzia, 225
Eschscholzia californica, 61, 67, 130, 170, 174, 225
    subsp. *mexicana*, 160, 163, 226
Esselen monkey flower, 255
establishment period, 296
Eucalyptus macrocarpa, 56, 226
Eucalyptus microtheca, 148
Euphorbia, 26, 227
    'Blackbird', 29, 75, 81, 228
Euphorbia ammak var. *variegata*, 44
Euphorbia antisyphilitica, 227
Euphorbia characias, 228–229
    'Tasmanian Tiger', 144, 146
Euphorbia cotinifolia, 44, 81, 118, 119, 122, 229
Euphorbia × Martinii 'Ascot Rainbow', 67, 230
Euphorbia mauritanica, 142
Euphorbia milii, 119, 230
    'Apricot', 74, 81
Euphorbia myrsinites, 135, 231

Euphorbia pulcherrima, 227
Euphorbia rigida, 19, 29, 53, 142, 154, 231
Euphorbia tirucalli
    'Flame', 89
    'Sticks on Fire', 42, 56, 57, 119, 122, 232
evergreen grasses, 301

fan aloe, 128, 142
feathery cassia, 206
Felicia fruticosa, 57
Ferocactus, 232
Ferocactus pilosus, 159, 163, 232–233
Ferocactus wislizeni, 161, 163, 233
fertilizing, 137, 302
Festuca californica 'Phil's Silver', 61
Festuca idahoensis
    'Siskiyou Blue', 61
    'Stony Creek', 61
Ficus repens, 107
Fielder, Mary Beth, 68, 72
firecracker penstemon, 263
firecracker plant, 75
fishhook barrel cactus, 161, 233
flapjack plant, 19, 241, 269
Flower Carpet roses, 266
flower dust plant, 242
foothill penstemon, 61, 263
Forestiera neomexicana, 133
Fouquieria splendens, 158, 159, 163, 234
foxtail fern, 103
Fragaria 'Pink Panda', 95
Frank, Gabriel, 25, 29, 97, 101, 122, 289
Frankenia thymifolia, 75
Fremontodendron californicum, 165
frosty blue California lilac, 210
Furcraea foetida 'Mediopicta', 235
fuzzy plants, 294

Gaillardia, 43
Gaillardia ×*grandiflora*, 235
    'Fanfarc Blaze', 100, 235
gayfeather, 252
geranium, 20, 128
Geranium 'Rozanne', 93, 94
germander sage, 152, 167, 269
giant Dutchman's pipe, 80
giant hen and chicks, 77, 220
Gimbel, Dustin, 29, 44, 73–81
Glandularia bipinnatifida, 151, 153
Glaucium flavum, 170
globe mallow, 15, 151
gold coin aster, 91
golden barrel cactus, 50, 170, 221
gopher spurge, 19, 53, 142, 231
gorilla's armpit, 161
Graptoveria 'Fred Ives', 29
grasses, 42, 112, 286, 301
grass tree, 213
gray water, 298
Grevillea, 302
    'Long John', 236
    'Moonlight', 123, 129
grevilleas, 301
Grinton, Phil, 58
Grisdale, Dinah, 18
grooming, 301–302
ground morning glory, 44

Heard, Mary Lou, 73
Hella Glashoff yarrow, 176
Heller's penstemon, 264
Hemerocallis 'Lady Eva', 91
Hesperaloe parviflora, 149, 236
Heteromeles arbutifolia, 65, 127
    'Davis Gold', 112
hill country penstemon, 161, 264
Hinkley, Dan, 73

Index

honeybush, 26, 168, 254–255
hopseed bush, 80
Hornberger, Michelle, 90–92, 95
horned poppy, 170
Horton, Judy, 21, 123–130
hot lips little leaf sage, 272
Hummel's jade plant, 217
hummingbirds, 93, 262, 268, 270, 273
hummingbird sage, 112, 273
Hunnemannia fumariifolia, 168, 171, 174, 237
hydrozones, 291–292
hypertufa, 77

Indian mallow, 57
in-line drip irrigation, 60, 87, 94, 105, 111, 172, 289–290
instant landscapes, 285
Iris, 238
    'Canyon Snow', 239
    'Clincher', 239
    'Dorthea's Ruby', 239
    Pacific Coast hybrid, 239
    'San Ardo', 239
Iris douglasiana, 239
Iris innominata, 239
irrigation, 60, 87–88, 94, 105, 111, 130, 137–138, 152, 172, 289–292, 296–298
island pink yarrow, 178

jelly bean monkey flowers, 256
Jerusalem sage, 19
jester conebush, 248
John Dourley manzanita, 196

Kalanchoe, 240
Kalanchoe fedtschenkoi, 240
    'Variegata', 240
Kalanchoe luciae, 19, 23, 89, 114, 118, 122, 146, 241, 269

Kalanchoe pumila, 242
Kalanchoe thyrsiflora, 241
kangaroo paw, 42, 91, 191–192, 299
Kniphofia, 44, 301
    'Malibu Yellow', 100, 101
Kniphofia thomsonii, 26

Lagerstroemia, 154, 243
landforms, 288–289
Lavandula, 244
Lavandula allardii 'Meerlo', 244
Lavandula angustifolia, 245
Lavandula ×intermedia 'Grosso', 95, 245
Lavandula latifolia, 245
Lavandula stoechas, 246
    'Wings of Night', 91, 95, 246
lavender, 150, 244–246
lavender sage, 136
lavender scallops, 240
lawn removal, 284–287
leaves, 294
Lerner, Cheryl K., 33, 35
Lester Rowntree manzanita, 197
Leucadendron, 29, 246, 302
    'Ebony', 168, 172, 174, 247
    'Jester', 100, 101, 105, 114, 119, 121, 122, 248
    'Safari Sunset', 124
    'Wilson's Wonder', 85
Leucadendron discolor 'Pom Pom', 247
Leucadendron salignum 'Winter Red', 89, 174, 248
Leucophyllum langmaniae
    'Lynn's Everblooming', 249
    'Lynn's Legacy', 137, 138, 249
Leucospermum, 119, 250, 302
    'Veldfire', 74, 81, 174, 251
Leucospermum cordifolium 'Yellow Bird', 174, 250
Leucospermum reflexum, 251
Liatris punctata, 138, 252

Index

Ligustrum sinense 'Variegatum', 123
lilac verbena, 278
Lima, Amelia, 24, 40, 82–88
lipstick echeveria, 219
Lloyd, Christopher, 54, 56, 73
Lobelia laxiflora, 22
Lonicera sempervirens 'Major Wheeler', 93
Loseley Park, 19
low-phosphorus fertilizers, 302
low-water gardens, 8–12
low-water landscapes, 285
Lychnis coronaria, 129
Lynn's legacy Texas sage, 249

Madagascar ocotillo, 121
Magnolia ×*soulangiana*, 95
Majorelle, Jacques, 126
Manfreda, 253
×*Mangave* 'Macho Mocha', 253
manzanita, 61, 194
Marx, Roberto Burle, 82–84
Mascagnia macroptera, 163, 254
matilija poppy, 166
McDougal, Carol Anne, 31
Mediterranean-climate gardens, 176
"Mediterranean mounds", 289
Meerlo lavender, 244
Melaleuca, 98
Melianthus major, 26, 168, 174, 254–255
    'Purple Haze', 27, 44, 77, 80, 81, 174, 255
mescal bean, 275
Mexican bush sage, 24, 93, 271
Mexican cardinal flower, 22
Mexican fan palm, 203
Mexican gold poppy, 160, 226
Mexican grass tree, 84
Mexican lily, 86, 200
Mexican lime cactus, 159, 232–233
Mexican orchid vine, 254

Mexican snowball, 219
Mexican tulip poppy, 168, 237
Mimulus, 255
    'Changeling', 61, 62, 67, 256
    'Jelly Bean Dark Pink', 256
    'Jelly Bean' hybrids, 72, 256
    'Jelly Bean Lemon', 256
    'Jelly Bean Orange', 256
    'Jelly Bean Purple', 256
    'Trish', 71
Mimulus bifidus 'Esselen', 62, 67, 255
monkey flower, 62, 71, 255–256
montbretia, 170
moonshine yarrow, 179
Moroccan daisy, 93
Morton, Laura, 20, 21, 106–114
mottlecah, 56, 226
mounds, 105, 288–289
mountain aloe, 56
Muhlenbergia, 301
Muhlenbergia capillaris, 65, 67, 173, 257
    'Regal Mist', 136, 257
mulch, 285, 299–302
multiple color schemes, 27
Mulvhill, Michael, 96–105
Mulvhill, Theresa, 96–97
mycorrhizae, 137
myrtle spurge, 135, 231

nasturtium, 19
native buckwheat, 63
Neomarica caerulea, 110, 258
Nepeta, 93
New Mexico, 176
New Mexico olive, 133
New Zealand flax, 86, 91, 213
noble aeonium, 181
nodding pincushion, 250
Nolina texana, 132

Index

ocotillo, 159, 234
Olsen, Les, 49
Opuntia, 147, 150
Opuntia macrocentra, 258
Opuntia violacea var. *macrocentra*, 160, 161, 163, 258
orange clivia, 212
organic fertilizers, 302
organic matter, 288
organic mulches (plant-based mulches), 299, 302
Origanum dictamnus, 43, 127, 128

paddle plant, 241
palo verde, 165, 259
paprika yarrow, 178
Parkinsonia 'Desert Museum', 22, 154, 165, 168, 259
Parkinsonia florida, 147
Pedilanthus bracteatus, 173, 260
Pedilanthus macrocarpus, 163, 260
pelargonium, 19
Pelargonium
    'Gary's Nebula', 126, 128
    'Veronica Contreras', 110, 111
Pelargonium sidoides, 93, 95, 261
pencil milk bush, 142
Peniocereus greggii, 154, 160
penstemon, 262–264
Penstemon, 262
    'Apple Blossom', 93
Penstemon ambiguus, 138, 262
Penstemon eatonii, 72, 263
Penstemon heterophyllus, 263
    'Margarita BOP', 61, 62, 72
Penstemon superbus, 160
Penstemon triflorus, 161, 163, 264
pepper tree, 173, 285, 302
Persian rockcress, 132, 135
Peruvian lily, 110, 190–191
Phacelia campanularia, 160

pheasant's tail grass, 98
Phillips, Judith, 131, 132–133
Phlomis fruticosa, 19
Phoenix robelini, 100
Phormium, 213
    'Bronze Baby', 91, 93
    'Dusky Chief', 86
phormiums, 299
phosphorus, 302
Pilosocereus pachycladus, 49, 50, 57, 264
pincushion, 74, 119, 250–251, 301
pineapple guava, 112
pink muhly grass, 65, 257
pink powder puff, 53
pink rockrose, 212
pink sugar African daisy, 199
Pittosporum tenuifolium
    'Silver Magic', 130
    'Silver Sheen', 130
planters, 92, 160–161
planting, 175, 292–296
poinsettias, 227
pointleaf manzanita, 197
pomegranate (edible and ornamental, dwarf and standard), 22, 265
pom pom conebush, 247
ponytail palm, 44, 119, 121
poppies, 71, 167, 225–226
pottery, 31–35
pozo blue sage, 272
prairie verbena, 151
prairie zinnia, 280
precipitation, 176
primary color schemes, 25
propeller plant, 216
Protea, 302
proteas, 301
pruning, 285, 301
pumpkin pie African daisy, 199
Punica, 22

Punica granatum, 130, 265
    'Nana', 65
purple lavandin, 245
purple rockrose, 211
pygmy date palm, 100

Quercus agrifolia, 285
Quercus fusiformis, 133

Radcliff, Doreen, 131–138
Radcliff, Phillip, 131
rain chains, 104
rainwater, 105, 298
red buckwheat, 223
red grass tree, 84
red hot poker, 44, 100
red pencil cactus, 232
red pencil tree, 56, 232
red yucca, 149, 236
Rhodanthemum hosmariense, 93
Richards, Alan, 31, 147–154
rocket pincushion, 251
rockrose, 135, 211–212
rocks, 12
Roger's California grape, 279
Romneya coulteri, 166
Rosa
    'Chocolate Sundae', 108, 110
    Flower Carpet, 67, 266
    'Joseph's Coat', 114
    'Veilchenblau', 111
rose, 18
Rose, Ingrid, 13–14
rose campion, 129
rosemary, 267
Rosmarinus officinalis, 67, 267
    'Arp', 267
    'Tuscan Blue', 267
Ruellia peninsularis, 154, 268
Russelia 'Night Lights Tangerine', 75

sage, 268–273
saguaro, 160
Salvia, 268
    'Bee's Bliss', 70, 72, 269
    'Plum Wine', 93
    'Pozo Blue', 72, 272
Salvia chamaedryoides, 22, 138, 152, 167, 174, 269
Salvia clevelandii, 70, 150, 269, 270
    'Compacta', 72
    'Winnifred Gilman', 114, 130
Salvia dorrii, 136, 137, 138, 270
Salvia greggii, 150, 153, 268
Salvia lavandulifolia, 136
Salvia leucantha, 95, 271
    'Midnight', 271
    'Santa Barbara', 271
Salvia leucophylla, 24, 93
Salvia microphylla
    'Hot Lips', 60, 61, 67, 272
    'Little Kiss', 272
Salvia sonomensis, 269
Salvia spathacea, 72, 112, 114, 273
sandpaper verbena, 278
sand penstemon, 262
sandy soil, 287
San Miguel island buckwheat, 223
Schinus molle, 50, 173, 285
Schumann, Judy, 155–162
sea heath, 75
sedge, 46
Sedum nussbaumerianum, 23, 49, 57, 74, 77, 85, 87, 89, 130, 273
Senecio mandraliscae, 26, 43, 85, 89, 119, 122, 127, 128, 130, 143, 144, 146, 170, 171, 274
Senecio serpens, 53, 87
Senna, 206
sentinel manzanita, 194–195
sheet mulch, 286
shoestring acacia, 150

Short, Terry, 90–92, 95
shrub aster, 57
Sierra gold dalea, 217
Sierra starr fairy duster, 204
silvery cassia, 207
single color schemes, 17–21
Sissinghurst Castle, 17–19
slipper plant, 260
"slow-release" fertilizers, 302
smoke tree, 110, 167, 214
Snowy River wattle, 112
soil, 287–289
solarization, 286–287
Solidago spathulata var. *nana*, 138, 274
Sophora, 148
Sophora secundiflora, 275
    'Silver Peso', 154, 159, 163, 275
    'Silver Sierra', 275
South African geranium, 93, 261
Spanish lavender, 91, 246
Spencer, Scott, 24, 166, 168
Sphaeralcea, 15, 151
Sphaeralcea ambigua, 174, 275
    'Louis Hamilton', 170, 172
spine-covered plants, 294
spring crocus, 61
spurge, 26
Stapelia grandiflora, 161
St. Catherine's lace, 129
Straw, Judy Mae, 139–144
strawberry tree, 192–193
streetside gardens, 15, 40
Strelitzia juncea, 50
strong contrasts color schemes, 28
succulents, 294
sundrop, 99, 167, 205, 269
sunset manzanita, 198
superb penstemon, 160
swales, 288–289

Tallamy, Douglas, 68
tall papyrus, 100
tall verbena, 277
Tecoma capensis, 107, 276
Tecomaria capensis, 276
Tecoma stans, 276
    'Bells of Fire', 276
    'Crimson Flare', 276
    'Sunrise', 276
terra cotta, 16–17
Teucrium fruticans, 125, 127
Texas mountain laurel, 275
Texas sage, 137
Thelocactus, 160, 161
tilt-head aloe, 102, 189
tone-on-tone color schemes, 21–24
toyon, 65, 112, 127, 129
Trainor, Bernard, 39
Trichocereus, 154
tuxedo spine prickly pear, 161, 258
twisted acacia, 150

unusual color combinations schemes, 27–28

variegated octopus agave, 186
variegated spurge, 228–229
veldfire pincushion, 251
verbena, 278–279
Verbena, 277
Verbena bonariensis, 277
Verbena lilacina 'De la Mina', 72, 278
Verbena rigida, 278
Vinatieri, Shelly, 115–117, 121
Vitis 'Roger's Red', 20, 111, 114, 126, 279

walking iris, 110, 258
warm color schemes, 24–25
water management, 296–298
waterwise plants, 294

Watsonia, 170, 173, 301
weed cloth, 299–301
weeping acacia, 73
western yarrow, 177
Whibley, Reg, 139–144
white-striped century plant, 85, 183
wildflower seeds, 160
wild strawberry, 95
Wilkinson, Nick, 42, 44, 115–122, 139–143
winter red conebush, 248
Wisteria 'Black Dragon', 114
wooly butterfly bush, 203
Wright's buckwheat, 224

yarrow, 71, 176
yellow bells, 276
yellow yucca, 236
Yucca, 213
    'Bright Star', 42, 57, 105, 146, 279
Yucca linearifolia, 102
Yucca rostrata, 102

Zauschneria californica, 222
Zauschneria canum, 222
Zinnia grandiflora, 135, 138, 280

© Mim Michelove

Nan Sterman is passionate about plants. With degrees in botany, biology, and education, she channeled her passion into a career as an award-winning journalist. Her work has appeared in major gardening publications on the regional, national, and international levels. Nan spends much of her time working on her Emmy award–winning TV show, *A Growing Passion*, which airs on public television in San Diego and is posted online at AGrowingPassion.com. Nan lives in Encinitas, California, where she speaks, teaches, and writes about low-water, sustainable, and edible gardening, and designs landscapes for client's homes and public spaces.